Advanced Introduction to Platform Economics

Elgar Advanced Introductions are stimulating and thoughtful introductions to major fields in the social sciences and law, expertly written by the world's leading scholars. Designed to be accessible yet rigorous, they offer concise and lucid surveys of the substantive and policy issues associated with discrete subject areas.

The aims of the series are two-fold: to pinpoint essential principles of a particular field, and to offer insights that stimulate critical thinking. By distilling the vast and often technical corpus of information on the subject into a concise and meaningful form, the books serve as accessible introductions for undergraduate and graduate students coming to the subject for the first time. Importantly, they also develop well-informed, nuanced critiques of the field that will challenge and extend the understanding of advanced students, scholars and policy-makers.

For a full list of titles in the series please see the back of the book. Recent titles in the series include:

Regional Innovation Systems
Bjørn T. Asheim, Arne Isaksen and Michaela Trippl

International Political Economy
Second Edition
Benjamin J. Cohen

International Tax Law
Second Edition
Reuven S. Avi-Yonah

Social Innovation
Frank Moulaert and Diana MacCallum

The Creative City
Charles Landry

International Trade Law
Michael J. Trebilcock and Joel Trachtman

European Union Law
Jacques Ziller

Planning Theory
Robert A. Beauregard

Tourism Destination Management
Chris Ryan

International Investment Law
August Reinisch

Sustainable Tourism
David Weaver

Austrian School of Economics
Second Edition
Randall G. Holcombe

U.S. Criminal Procedure
Christopher Slobogin

Platform Economics
Robin Mansell and W. Edward Steinmueller

Public Finance
Vito Tanzi

Advanced Introduction to

Platform Economics

ROBIN MANSELL

Professor, Department of Media and Communications, London School of Economics and Political Science, UK

W. EDWARD STEINMUELLER

R. M. Philips Professor of the Economics of Innovation, Science Policy Research Unit, University of Sussex, UK

Elgar Advanced Introductions

Edward Elgar PUBLISHING

Cheltenham, UK • Northampton, MA, USA

Published by
Edward Elgar Publishing Limited
The Lypiatts
15 Lansdown Road
Cheltenham
Glos GL50 2JA
UK

Edward Elgar Publishing, Inc.
William Pratt House
9 Dewey Court
Northampton
Massachusetts 01060
USA

A catalogue record for this book
is available from the British Library

Library of Congress Control Number: 2020940520

ISBN 978 1 78990 060 6 (cased)
ISBN 978 1 78990 061 3 (eBook)
ISBN 978 1 78990 062 0 (paperback)

Printed and bound in Great Britain by TJ International Ltd, Padstow, Cornwall

Contents

Preface

This introductory volume is intended for readers with and without formal economics training. Commercial platform operations, the collection and uses of data, and the huge scale of operations of some companies are often taken as inevitabilities. Innovation in artificial intelligence-enabled technologies is deemed essential to compete successfully on a global stage. Why?

Digital platform operations are intimately linked with changes in economies, in polities, in sociability and in democracies. They are tied directly and indirectly to inequalities, exclusions and disadvantage. In our view, these consequences are inevitable only when we fail to recognise why corporate and policy choices are being made and how they might be imagined and enacted differently. Economic analysis is used both to defend and to criticise current platform developments.

In this book we surface the assumptions and strengths and limitations of the dominant neoclassical economics tradition as well as two other economics traditions. The origins and consequences of digital platforms differ depending on which analytical lens is used. Our aim is to help readers to grasp these differences in a way that can potentially yield alternative visions for future platform developments.

Acknowledgements

Ed Steinmueller is grateful to Juan Mateos-Garcia for numerous helpful observations on platform developments including mitigation measures. Robin Mansell is grateful to Bingchun Meng for assistance and comments on our discussion about Chinese platforms and strategies, to Fabien Cante for literature review assistance, and to her students who have helped to sharpen her understanding of how economic theories are invoked to explain digital "disruption" in their countries. We are both grateful for the support of our colleagues at our respective universities and for time to write. We are hugely grateful to Professor Johannes Bauer for his generosity in making very constructive comments on an earlier draft. Finally, we are responsible for any errors or omissions.

Robin Mansell and Ed Steinmueller
London, January 2020

1. Introduction

Companies such as Amazon, Apple, Facebook and Google,[1] as well as Alibaba and Tencent, have experienced phenomenal economic performance. Many other companies offering online services are expanding rapidly, sometimes through profitless growth. These companies are sources of both wonder and consternation. Known as digital platforms, they are a striking 21st-century development. They are not just novelties. They are delivering benefits in the economic, political, social and cultural realms. At the same time, their business operations are widely believed to be infringing upon people's fundamental rights such as the right to individual privacy and the right to freedom of expression. In this book, we hope you will find answers to questions about the benefits, costs and hazards that digital platforms are implicated in creating. You will acquire a working knowledge of how several traditions in economic analysis address how and why digital platforms are influencing every facet of society. We also examine options to influence platform market competition and we discuss how platforms might better engage with public values.

1.1   What is novel about digital platforms?

Digital platforms are unprecedented. Never has it been possible to gather or process as much data about individuals' choices, behaviours and characteristics or to link these to their potential interests, buying preferences and worldviews. Modern digital platforms are distinguished by their use of technologies for linking multiple suppliers and consumers or citizens. They establish these links by making use of data gathered either directly from users or by observing their behaviour. The commercial platforms collect massive amounts of data. This data is monetised, typically, through the sale of marketing and advertising services. This means of generating revenue – or business model – figures centrally in many peoples' lives

1

around the world due either to their participation on digital platforms or to their exclusion. The attention of platform users is being directed and shaped in the interests of making money. The potential for generating revenue is creating strong incentives to devise attractions for platform users, even when those attractions are morally or legally dubious.

Platform owners are intent on developing their capacities to predict attention. They seek innovative ways to intensify their bids for user attention – all the while improving the targeting of users for advertising – commercial and political. The economic outcomes of these processes for the platforms are higher revenues and the removal of oxygen (advertising revenue) from rival sources of attention (e.g. the traditional media). The resulting digital environment offers substantial benefits from access to information, wider choice and desired social connections. It is also associated with risks and harms with the potential to damage cultures and democracy.[2] These benefits and costs are linked to economic value and public values.

Whereas economic value is about wealth creation and distribution, the term "public values" may refer to issues of fairness, equality and both individual and collective solidarity. In the discussion of platforms, among the public values of interest are fundamental rights to freedom of expression, individual privacy and freedom from monitoring or surveillance (van Dijck et al., 2018). These values, in turn, are affected by the platforms' influence on the desirability and provision of diverse information in the public sphere. All these values exist in tension both with each other and with the desirability of generating economic value. It is for this reason that the platform phenomenon is sparking intense policy debate about how their operations should be governed. There is limited scope in this book to examine how the relative importance of private or economic value and public values might be decided in decision-making processes. However, since traditions in economics make different assumptions about this crucial matter, we consider how they influence efforts to achieve platform accountability and socially acceptable approaches to platform governance.

1.1.1   Digitalisation

The rise of the technologies underlying digital platforms has been underway for several decades. Different choices might have been made concerning the pathway of their development. A step that was essential for their development was the digitalisation of the communication infrastructure.[3] Digitalisation involved a large increase in the volume of data of all types that is gathered, the capabilities for analysing data and the purposes for which it may be used. It resulted in data resources that could be used to extract, process and create economic or other kinds of value. Innovations that enabled digitised information to travel through a communication network occurred in the 1950s and 1960s. By the early 1970s, computerisation was enabling "a storehouse of virtually untapped new and improved services to the public".[4] By the 1980s, new digital telecommunication capabilities emerged (e.g. Calling Line Identification which displayed a caller's number on the recipient's phone). The new technical capabilities were applied initially for customer billing, identifying sources of emergency calls and tracing malicious calls. As these capabilities were developed, they raised concerns about the surveillance by companies and governments of the private spaces and interations of individuals. Privacy and civil liberty issues became prominent subjects for policy debate (Mansell, 1996; Samarajiva and Mukherjee, 1991). It was recognised that behaviours and practices, outside the usual social norms about privacy and surveillance, were being fostered by these developments; and the ability to collect and process data was characterised as an all-seeing "electronic eye" (Lyon, 1994; Zuboff, 1988). The application of these systems to enable discriminatory "social sorting" of groups or classes was also noted (Gandy, 1993).

During the 1990s, the internet emerged with global data communication capability. Its neutrality in serving all information producers and users was promoted as upholding a right to freedom of expression, and, in some countries, with the prospects of a libertarian bypass of social norms and government control. The internet's openness would, however, be contested (van Schewick, 2010; Zittrain, 2003). For better or worse, the platform services built upon this network infrastructure and further advances in digitalisation came to be regarded as a major "driver" of economic growth in what has come to be known as the "digital economy".[5]

1.1.2   Mobility

Innovations in mobile technologies freed users from fixed terminal nodes of the network (Katz and Aakhus, 2002). Cellular (or mobile) communication was invented in the United States (US) in the late 1940s, although radio communication has a much longer history (Raboy, 2016). The first handheld mobile phone was available in the early 1970s, smartphones from the mid-1990s (IBM Simon, BlackBerry) and Apple's iPhone and its App Store from 2007 and 2008. Users, untethered from their desktop computers, now had even greater abilities to create and circulate text, audio and visual content.

By the late 1990s, mobile service providers had started to adopt a platform model integrating the power of computing and software in their networks (Ballon, 2009; Bresnahan and Greenstein, 2014), and building on the spread of the internet to offer early mobile media services (Funk, 2003; Ibrus, 2010). These platforms began to be used for the distribution of both professionally produced and user-generated content. Continuing improvements in connectivity, combined with the linking of businesses and individuals to a mobile and increasingly accessible internet, encouraged the development of cloud-based services, such as those offered by Amazon Web Services, Microsoft Azure and Google Cloud (Mosco, 2014). Faster next generation mobile networks (5G) promise to carry enormous quantities of data captured from human communication and from communication among things – the Internet-of-Things (IoT) (which had been discussed earlier as ubiquitous or ambient computing). The IoT uses sensors embedded in physical objects and connected to the internet. It enables novel applications, including autonomous vehicles, smart cities, new health services and innovative retail systems, as well as wearables such as health monitors. Like previous developments in digitalisation, IoT components are becoming part of everyday life and are invisible, even while they vastly extend the ability to track, observe and anticipate people's activities.[6]

1.1.3   Datafication and artificial intelligence

Digitalisation greatly augments the means for creating value from the collection and processing of data. This value need not be directly monetary.[7] For instance, data about patients or students can deliver social benefit through improved health or learning. For much of the 20th century, developing means for collecting data to improve control was a major business activity. Companies built customer loyalty based on information about customer shopping habits long before the internet, but the computerisation of networks allowed them to manipulate and use data in new ways. Multimedia applications in the 1990s, alongside data communication services (including e-commerce) and digital content started to be treated as major "drivers" of economic opportunity in a dawning information age. By the end of the 1990s, as early platform-like services developed, conflicts between customer or citizen and advertiser interests were recognised. Negotiating these conflicts became a priority for companies that needed to retain user interest and loyalty while preserving advertiser support. Earlier, some had hoped that computerised systems embedded in networks would be "programmed so that *users dictate* the nature and extent of computer processing applications".[8] During the first decade of the 21st century, this expectation evaporated. Platforms as well as other parts of the online world were operating in ways that were beyond the knowledge or consent of their users.

A further phase of innovation in the form of artificial intelligence (AI) and machine learning is now underway. Substantial changes in user online interaction practices are complicating the boundaries between public and private life. The platforms are implicated in the reshaping of civic and political discourse, the dissipation of social cohesion and, for some, the enslavement of human learning and imagination (Zuboff, 2019).

The platforms' uses of data derived from user monitoring are what van Dijck and others have designated as datafication: the use of data gathered about users (van Dijck et al., 2018). Datafication involves converting any phenomenon, including user behaviour occurring during online interactions, into data: "a quantified format so it can be tabulated and analysed" (Mayer-Schönberger and Cukier, 2013: 78). Datafication, especially when undertaken by very large commercial platforms, raises concerns about the consequences when platform operations transform "online and offline objects, activities, emotions, and ideas into tradable commodities" (van Dijck et al., 2018: 38).[9] Datafication may be deployed for functions ranging from price discrimination to user profiling. Advertising may be bundled with content generated by users, e.g. social media, or it may be placed by platform owners, e.g. in search results. User monitoring or surveillance is allowing (largely) automated choices about the placement of advertising content. The aim is to increase the likelihood that platform users will engage with it. Datafication means that economic value can be generated by a platform merely by capturing data about user interests and preferences as revealed by their interactions on the platform.

1.2   Defining digital platforms

The term "platform" in this context is a metaphor and it has multiple meanings (Gillespie, 2010). In the economics and management literatures, the modularity of platform components and strategies for combining them into ecosystems are key concerns and a platform is defined as the "systematic re-use of components across different products within a product family" (Gawer, 2014: 1242). The analytical focus is often on platforms as intermediaries which aim to attract customers who seek to interact on beneficial terms (Evans and Schmalensee, 2016). An intermediary is a third party to a basic exchange between a producer and a consumer (or individuals who wish to communicate for social or political purposes) (Hagiu, 2007; Rochet and Tirole, 2003). A digital platform business model defines the mix of direct and intermediate services that a platform offers. Digital platforms differ, however, from the basic intermediary model in two ways. First, they retain information about customer purchases *and* about their patterns of online behaviour that may or may not lead to a purchase. The second distinction concerns the way platform owners use customer or user-created content to earn revenues taking advantage of network effects.

The definition of what a digital platform is depends on what we want to know about it and there are many ways of defining and classifying platforms (see Gawer, 2011; Lehr et al., 2019; Nooren et al., 2018). Our interest is in how platform operations are related to both economic value and a range of public values. To develop a definition of platforms for this purpose, we begin by following van Dijck et al. (2018: 4) who characterise a platform as "a programmable digital architecture designed to organise interactions between users – not just end users but also corporate entities and public bodies". Platforms also engage in the development of ecosystems or "an assemblage of networked platforms, governed by a particular set of mechanisms ... that shapes everyday practices" (van Dijck et al., 2018: 4). Our working definition has four elements: (1) content desired by users; (2) a business model that pays the costs of maintaining and improving the platform; (3) the collection, retention and use of data about users; and (4) the provision of auxiliary services.

These elements are explained in detail in Chapter 2. The first three elements are essential. The fourth is prevalent, though not universal, and often involves use of data about users. Our working definition is intended to encompass both commercial platforms and those without the intent of accumulating profits. It excludes the many websites that gather data from or about users but which do not make systematic use of that data to shape or reinforce user attention. This definition is inclusive in that it does not specify the nature of the business model employed other than that it provides the means to maintain and improve the platform. Other authors (such as Hagiu and Wright, 2015a, 2015b) introduce further distinctions that seek to differentiate platforms from other online activities based upon whether the platform owner is able to participate simultaneously in (and exercise some control over) multiple markets, which we consider further in Chapter 3 (3.2.1). These further distinctions are useful for questions involving market coordination, an important business model, but they neglect other business models based on the simpler principle of monetising or deriving non-monetised social value from user data and the observation of users.

1.3   Platform consequences

Datafication practices which exploit user attention are central to the leading commercial social media and e-commerce platforms. These platform developments generate different responses depending on which economic lens informs analysis of opportunities and pitfalls. In this book we focus on different, but sometimes overlapping, traditions in economic analysis – *neoclassical economics, institutional economics* (sometimes designated as political economy) and *critical political economy* (inspired by Marxist traditions). Each offers insight into the digital platforms' rise to prominence, their consequences and, in some instances, what can and should be done to govern them. We pay attention, particularly, to the norms and rules that inform market, civil society organisation, state and individual practice in relation to these platform developments. We focus especially on norms and rules because the way these are institutionalised is inseparable from the way markets work (Freeman, 1988).

Policy makers often turn to economic analysis to inform their decisions about how to govern or regulate the digital platforms. Sometimes platforms are presented positively as empowering individuals and communities, but with little regard for the specific nature of individual or institutional practices or for the agency or consequences of the power that platforms exercise. At other times, the risks of, and actual, harms associated with commercial platform operations are emphasised. The platforms' operations are of special concern partly because of the increasing invisibility of their operations and partly because of the way they are seen to challenge the "settlements of constitutional democracies, namely democracy, rule of law and fundamental rights".[10] If digital technologies and platforms are to provide a societal foundation or infrastructure embodying human rights that yields fair and equitable outcomes, platform governance norms and rules consistent with these goals need to be in place. In the 2020s, platform governance in the Western world is likely to be deeply unsettled and this motivates our consideration in Chapters 5 and 6 of self-regulatory approaches and external policy and regulatory interventions. Questions about policy and regulation in Western regions and countries are also contested when they present themselves in relation to choices taken in other regions of the world, which we consider in Chapter 7.

1.3.1   Technologies and societal change

The technical, architectural and organisational features of digital platforms place them at the centre of economies in which the collection, processing and interpretation of data hold enormous promise for wealth creation and improvements in human welfare. In one view, this is because technological innovations are expected to provide the foundations for capitalist economies that guarantee new jobs and greater consumer choice. In this view, digital platforms are said to benefit from the "disruptive" characteristics of the Fourth Industrial Revolution (Schwab, 2017). Disruptive technologies – computer hardware and software, AI and machine learning – are regarded as impacting on society. The challenge is to *adjust* market and societal norms and rules to accommodate the characteristics of these technologies. The term "disruptive" is also used in the business literature to refer to strategies capable of destabilising the established positions of incumbent firms.[11] Such strategies include changes in the design of products that allow them to be produced more cheaply but that are difficult for incumbents to reproduce. This is because they draw upon a different knowledge base, e.g. the introduction of digital copying as a substitute for xerography in photocopier machines.

There are numerous ways in which economic value generation and public values could be organised around digital technologies. The notion of "adjustment" is far more complicated than the standard accounts which focus principally on novel technologies. "Disruption" is viewed in this more complicated context as a process with multiple causes, only some of which are linked to digitalisation and datafication. Processes of adjustment are informed by power relations among multiple actors, relations that are often asymmetrical in practice. In this sense, digital platform companies can be characterised as disruptive innovators insofar as they build innovative business processes that existing companies must react to if they are to survive. They can be "disruptive" when consumers' or citizens' views or behaviours are influenced by using them or by others' use of them. "Adaptations" associated with these developments alter societal norms and rules in minor or radical ways. Some adaptations may deepen areas of concern; others may favour alternative approaches for platform operations with greater potentials for public values to be upheld.

We use the term disruptive broadly to refer not only to deliberate strategic choices of businesses, but also to the intersections between the design and deployment of new technologies and established rules, norms and standards – including cultural, social and political as well as economic institutions and practices. Our usage of the notion of disruption is similar to Schumpeter's evocation of "creative destruction", a process of change that may begin with innovation but overturns and creates new societal orderings. In this view, risks and harms associated with digital platforms are not in any sense determined by technological innovation: "technological progress is not a force of nature but reflects social and economic decisions" (Atkinson, 2015: 3). Our analysis of innovation processes focuses on how technology and society influence each other – in economic analysis, how "a better match between the new technology and the system of social management of the economy" can be achieved (Freeman and Perez, 1988: 38).

1.3.2   Challenges to the social order

The operations of the digital platforms are sparking debate within countries and across the world because of concerns that they are diminishing the capacity to ensure that public values are upheld. Critics, especially of dominant commercial platforms, argue they are complicit in promoting surveillance capitalism which aims to modify individual behaviour for profit and by stealth (Zuboff, 2019). Platform practices are being characterised as "data colonialism" (Couldry and Mejias, 2019) and as normalising a culture of surveillance (Lyon, 2018). Judgements about the risks and harms associated with their practices vary in severity in different contexts making it extremely difficult, so far, to introduce changes in norms and rules that the platforms have elected for themselves.

Platforms are operated on the assumption that consumers and citizens want to see personalised or targeted content and messages. Accomplishing this aim requires compromising traditional public-private boundaries to support the commercial platforms' business models that rely on data to feed their AI-enabled machine learning systems. Platform business models and practices target users based on increasingly granular computations of their demographics, interests, social connections, newsfeeds and mobility. These processes enable individual preferences to be interpreted using automated (algorithm) systems that yield predictions of behaviour. These predictions are being applied in commercial and political contexts with the expectation of improved decision making. Although in practice these predictive systems only offer a veneer of certainty, they are used to guide sellers' decisions about who is a likely purchaser of goods and services, who is likely to vote for a candidate in an election, or who is a likely criminal suspect.

Among consumers and citizens, there is concern about changes in the boundaries between public and private life and especially about individual privacy (Turow, 2011). Direct causal links between people's online interactions and online and offline harms are difficult to confirm, but there is broad agreement that people benefit from a degree of privacy and a sense of safety and security (Stoilova et al., 2019). These systems are enabling discrimination through advances in algorithmic techniques using training data that reflect pre-existing economic, racial, ethnic, gender and other discriminations that exist within societies: the result is the "automation of inequality" (Eubanks, 2018).

The amplification of the availability of illegal and harmful digital content which has accompanied the rise of platforms is coinciding with declining abilities to assess the accuracy of online information. Characterised as an information crisis[12] or as a democratic crisis of representation (Pariser, 2011), the sustainability of a democratic public sphere in Western societies is being questioned and it is being argued that "we are now at a crucial juncture where trust and confidence in the integrity of our democratic process risks being undermined".[13] When measures aim to make platform operations more transparent and accountable to the public, however, the risk is that the right to freedom of expression will be jeopardised by over-zealous or overcautious censoring by platform owners.

Platforms such as Google and Facebook have strong incentives to win viewer attention to the exclusion of consideration of what it is that is attracting attention, because their priority is securing economic value from data. Attention can be gained by any number of emotive or visceral representational texts or images or with texts and images produced with an intent to accurately inform or educate an audience. The move towards news as entertainment in the last decades of the 20th century was a precursor to today's platform practices and a growing dependence of professional news production organisations on platform distribution. This dependence arises from the largest platforms' dominance over revenue sharing with the news publishers whose content they host. The deconstruction of editorial choice threatens the diversity of news content and the credibility of information that is available to platform users.

The digital platforms are also implicated in enlarging the ranks of those living in a precarious position with further challenges to social cohesion. Inability to sustain a living wage in the face of changes in workforce skill requirements associated with digitisation, and declines in workplace conditions as the digital platforms facilitate flexible and heavily routinised employment, reduce employment security. Together with constraints on entitlements to work due to immigrant or refugee legal status within countries, people are living with a declining sense of predictability in their lives; they are living more precariously. This is exacerbated when decisions affecting consumers and citizens are informed by opaque algorithms and when people have limited resources to contest interpretations of data that affect their lives. The unprecedented scale and reach of the digital platforms' operations, the absence of viable opt-outs for consumers and citizens, and the biased outcomes associated with their management and use of data mean that the platforms are deeply implicated in reinforcing societal inequalities and injustices, just as they are in generating economic wealth.

1.4   Structure of the book

Efforts to mitigate or avert risks and harms associated with digital platforms confront the argument that digitalisation and datafication are forces for good in society – we must adjust to the inevitability of technological progress. The counterargument is that these developments involve social processes and choices that change societal norms, rules and expectations – they therefore should be socially deliberated. The chapters in this book highlight the strengths and the limitations of the three traditions in economic analysis in understanding platform developments and for addressing the deliberations accompanying these developments.

In Chapter 2, we explain that a digital platform is best understood as a radical innovation with implications for the way economic value is generated and public values are upheld. By examining the elements of platforms and the complementary developments that aided their growth, we develop the working definition of platforms used in this book. We explain the limits of a purely market-oriented economic analysis for understanding these developments.

Chapter 3 develops the three economic frameworks for examining digital platforms by identifying their assumptions and focus of attention. We show how each framework contributes to an understanding of platform development.

Chapter 4 looks at the emergence of AI as an amplifier of the datafication process underlying platform developments and as a future source of market strength for platform companies. This chapter also identifies platform provision models that provide alternatives to the currently dominant advertiser-supported model.

Chapter 5 considers some of the rationales for and shortcomings of digital platform self-regulation and the potential for alternatives to the commercial datafication-driven business models to challenge the dominance of the advertiser-supported business model.

Chapter 6 explores external policy and regulatory approaches aimed at creating incentives for change in existing platform behaviour or at changing the boundary between public and private supply through structural measures.

Although our primary focus is on digital platform developments in the West, in Chapter 7 we consider global developments, including those in China, in the light of arguments about "catching up" with countries in the global North.

Finally, Chapter 8 summarises what we have learned about digital platforms and their consequences, highlighting choices that will influence future developments.

Notes

1. In November 2019, FACEBOOK was adopted as the corporate logo to distinguish the company from the Facebook app. Alphabet is the corporate owner of Google.
2. See O'Neill (2002), and for a list of harms see UK (2019e: 25).
3. Digitalisation refers to the practice of collecting, storing, curating and communicating digitally (binary 0s and 1s).
4. See Strassburg (1970: 12).
5. There is no fixed definition of the digital economy. UNCTAD distinguishes a core sector (hardware manufacture, software and information technology consulting, information services and telecommunications) and a narrowly defined "digital economy" including the platform economy and digital services, and a broader digitalised economy including e-business, e-commerce, industry 4.0, precision agriculture, algorithmic economy, sharing economy and gig economy (UNCTAD, 2019).
6. See Ren et al. (2019) for a detailed analysis of data sharing by IoT consumer devices in the UK and the US.
7. Data can be treated as intangible assets because data holdings may elevate the market value of a company beyond its tangible asset value (assuming corporate ownership rights are in place). Assuming such rights, the *asset value* of data is like a natural resource (e.g. oil or fresh water), but this value is only a potential for generating revenue until it is employed for some revenue-generating activity such as the sale of advertising or the provision of market research services.
8. See FCC (1980: para 118), emphasis added.
9. Datafication processes are also discussed in relation to "platformisation" (van Dijck et al., 2018).
10. Bayamlioglu et al. (2018: 1) quoting Paul Nemitz, Principal Advisor, European Commission Justice Directorate General.
11. The use of "disruption" in the business strategy sense was introduced by Christensen (1997) and subsequently broadened in ways that obscured the original intent. See Gans (2016) for a useful restatement and revision of the basic theory.
12. See Trust Truth and Technology Commission (2018).
13. See UK (2018c: 47).

2. Digital platform origins and novelty

2.1   Introduction

We examine the origins of digital platforms in this chapter with the aim of creating a working definition of a platform. Many of the larger consequences of digital platforms arise from their novelty and their dynamically changing character. We draw upon the innovation literature to examine platforms as a means of organising markets. This also provides a framework for understanding societal changes accompanying digitalisation and datafication as a multifaceted process (as discussed in Chapter 1). We identify how choices taken at the origins of digital platforms fit together to define a novel structure of organisation where novelty lies, not in any single element, but in their combination. This combination is a radical innovation, an innovation with broad economic and social consequences because platforms are affecting the lives of a large and growing share of the world's population. The chapter also identifies norms, rules and standards that enable the common platform business models to emerge and flourish.

2.2   Digital platforms as radical innovations

Digital platforms are a rare instance of a new thing under the sun; what innovation scholars call a radical innovation (Freeman and Perez, 1988). Radical innovations such as steam power, electricity and the automobile enabled transformations in people's lives and relationships. An innovation is radical if it is a fundamental departure from established methods or processes and is associated with major changes in society, including social, political and economic relationships. Radical innovations are uncommon because most efforts to improve technology are focused on incremental change, e.g. product improvements and cost-saving alterations in design or production.

The outcomes of a radical innovation are neither pre-ordained nor inevitable, even if, in retrospect, they may seem to be when viewed through a narrow analytical lens. Multiple decisions inform the framework of norms and rules that open opportunities and impose constraints for technological innovation and its application. Thus, choices by companies, individuals, civil society groups and governments influence how a technology develops. Close examination typically reveals that it is a series of smaller and larger changes that channel or shape the direction of a radical innovation. Without these changes, innovations that come to be recognised as radical would be curious novelties; steam power was initially used to pump water out of mines, electricity was a laboratory curiosity, and the automobile was a hobbyist accessory for the rich. Building upon digitalisation and datafication processes as discussed in the previous chapter, two clusters of innovations in data communication networks – the internet and common standards for the World Wide Web – opened the door for the creation of digital platforms.

2.2.1 Establishing connectivity

Platforms emerged from commercial efforts to exploit fixed, and later mobile or wireless, broadband connectivity (Benkler, 2006; Lemstra and Melody, 2014), and with the flexibility of the internet following the end of commercial restrictions on the internet's use in 1995. By that year, a first cluster of innovations that would enable the platform innovation had already been developed. A diverse collection of data communication networks had been established under public authority and through private ventures during the previous 20 years. In order to interconnect them, protocols (a set of rules and technical standards) were devised to connect them. That this complex of networks converged to *the* internet is an instance of Albert Einstein's injunction (initially about scientific theory) – it should be as simple as possible, but not simpler. The internet came to be regarded as singular and to dominate as a means of data communication partly because of the simplicity of its standards and its layered architecture (Clark, 2018; van Schewick, 2010), although political factors played, and continue to play, an important role (DeNardis, 2014; Mueller, 2010). This simplicity had consequences for the spread of the internet, and the dominance of this set of rules and standards is implicated in some of the negative social, political and economic consequences of platforms.

2.2.2 Common Web standards

The second cluster of innovations which made digital platforms possible started with the development of the World Wide Web between 1989 and 1991. The Web operates with an additional set of rules and standards that creates an "information space" in which online resources (e.g. text, graphics and audiovisual content) can be accessed using an addressing scheme called the URL (Uniform Resource Locator). The URL can be thought of as a phone number or television channel allowing an internet user to connect to resources. This connection is often achieved using a web browser such as Microsoft's Internet Explorer, Mozilla's Firefox or Google's Chrome but also may be made with other software operating with or without user knowledge and control. The now familiar language of websites and pages refers to resources that browsers can access while other online resources are accessed directly using internet communication protocols. These resources are stored on a global network of servers. These servers are computers that combine elements such as content and advertising and receive and respond to inputs from users or from autonomously operating software. Just as internet standards have implications for social, political and economic developments in society, for both good and ill, so too do the standards for the Web, servers and web browsers.

2.3   Virtualisation of publication and commerce

These two clusters of innovations – internet connectivity and common standards for the Web – opened a space for a new medium of communication and for the emergence of digital platforms. They allowed publishing of information for and by a large and growing audience and they enabled the owners of websites to interact and monitor those accessing their resources.[1] They also, and crucially, provided a communication infrastructure linking software and computational processes to achieve digitalisation and datafication on a grand scale. The availability of large-scale or mass two-way communication – what would come to called mass self-communication (Castells, 2009) – was the key technical enabler making it possible to build digital platforms. The large-scale digital platform as we understand it today, however, was still only a potential.

A succession of developments would have to occur before the present-day digital platform operating in a complex ecosystem would appear. These developments were the result of three main trends: (1) the increasing and expanding capabilities of machines; (2) the market entry of companies bearing little resemblance to incumbent companies; and (3) the emergence of the globally distributed crowd, that is, a large global user community (McAfee and Brynjolfsson, 2017). They occurred through entrepreneurial experimentation by many actors – some public, some private – during the last half of the 1990s. Many of these efforts involved translating products and services into the online world created by the internet and the Web from the "brick and mortar" world. These efforts would involve deriving revenue and profit from datafication, not just of online activity, but also of emotions and ideas, rendering them as tradable commodities with major societal ramifications.

2.3.1   Offline to online translation process

The *first* and most obvious offline to online translation project was to devise websites that would reproduce the functions of print publishing – newspapers, magazines and books. The digitisation of previously published works began in 1971 with Project Gutenberg. This was an effort to create a shared library of works that were no longer subject to copyright and was undertaken in anticipation of broader public access to information resources long before it happened. Copyright was the basis of business models built on exclusive rights to publish. Translating traditional print publishing onto the Web is an example of a process that gave rise to the need for a new business model. The information resources provided by websites are, in principle, freely available to anyone who has the website address. The earlier business model – using copyright protection as the basis for enabling revenue to be gained from publishing (and excluding others from republishing) – no longer worked in the online world. It would not be possible to sell copies of published works without further innovation. Even with a simple innovation, the pay wall, a requirement to enter a name and payment information to gain access to published content, it was still possible that the work would be copied and shared with others. The inability to reliably use a previous business model that applied to copyright protected works gave rise to renewed efforts to control and manage online rights to access published information in the commercial market (Branscomb, 1994).

In addition to selling copies of copyright protected published works, another business model was available for use in the online world – advertiser-supported content. In this model, the user of information receives copies for free, accompanied by advertising messages paid for by advertisers. This model was well established in the newspaper industry where free distribution newspapers were profitably published with the aid of advertiser support. Advertiser-supported publishing would be adopted by many operators of digital platforms. A problem with this business model is that advertisers pay only a modest amount for an advertising message, typically a fraction of a penny for every viewer who is expected to be exposed to the message. This means that the online platform must attract many viewers to cover its costs and faces competition from others with advertising messages. In publishing, competition occurs among firms with different capabilities and values, raising many questions about access to and control of content.

2.3.2   E-commerce and online sales

A *second*, now-obvious application for the Web was the creation of online sales outlets for physical (non-digital) goods and services, e-commerce. E-commerce, as a business model, initially seemed a simple extension of earlier means of advertising and selling. For example, television channels devoted to infomercials for the latest kitchen and lifestyle gadgets tied to call centres accepting customer orders, or mail or postal order companies distributing catalogues through the post and receiving orders by post or phone. Amazon is a remarkably successful example of the translation of the advertise and order business model. Initially offering books and compact disc recordings for postal delivery, incorporated in 1994, it began in 1995 to function as an e-commerce site, one of many following the opening of the internet to commercial use. The ambitions of Jeff Bezos, Amazon's founder, may have been larger than many other e-commerce entrepreneurs of the time, but there was little reason to expect that Amazon would be the forerunner of a radical innovation. Amazon and other e-commerce sites did, however, demonstrate that the Web-empowered internet allowed for the creation of a new type of intermediary. These intermediaries would bypass the brick and mortar shops to offer an online shopping experience where product placement and customer service, including the information needed to support it, could be translated from previous retail experience.

2.3.3 Online search and indexing

A *third* stream of developments in business models in the 1990s is represented by Google. Unlike the two previous examples, neither the nature of the service nor the business model of Google's search engine was an immediate translation of earlier practice.[2] Google sought to create the best index of the Web by displaying a list of sites in response to a user's input of search terms. The list was ordered by the frequency with which a site was referenced by other sites and thus its popularity or significance. Google was neither the first nor the only search engine. However, its method of indexing (which continued to be elaborated) was innovative. Of substantial importance for its capacity to generate revenue, it adopted the advertiser-supported business model within two years of its launch. This allowed companies to pay for listing in the results displayed for the user-chosen search terms regardless of the popularity of the advertiser's site. Advertising-"supported" sites were indicated clearly in the list of search results initially, a distinction that would become less apparent over time.

What would be crucially important was Google's choice to record user behaviour in the use of its search engine and its decision to offer advertisers and others information about its users. The design of the Web and the web server software facilitated these choices because it was possible to observe and record user access to individual pages. This information proved valuable to advertisers, providing them with an incentive to embed code on their websites that would generate still more information that would be curated by Google. Other website controlling companies – Facebook and Amazon – also began to record and analyse data about their website users, but Google was one of the first to make extensive use of data generated from sites not under its direct control. These services would be taken up and extended as the commercial datafication business model. Central to the economic success of this business model was Google's default assumption that it could make use of the data that users generated through their interactions with the search engine or with other sites whose embedded code gathered data on user access and transmitted it to Google.

2.4   Working definition of a digital platform

What are the features that transformed content publishing, e-commerce, search engine and social media sites such iTunes, Amazon and Google or Facebook, as well as sites like Alibaba and Baidu, into what are now designated as digital platforms? Our working definition of this radical innovation has four components.

First, for a website to become a digital platform, it must have *content that makes it desirable for users to connect to it*. This is most obviously and directly fulfilled by publishing platforms, but users can also be attracted to content about goods and services on offer (e.g. Amazon) or content that provides a guide to potentially useful information (e.g. Google), as well as content professionally produced or produced by users that is deemed desirable by users (e.g. Facebook).

Second, the platform provider needs to have *a business model that will allow it to cover the costs* of creating or acquiring content and the software engineering and web page hosting services necessary to maintain the platform. In the case of publishers, this is the simultaneous delivery of advertising and content, with the former paying the bills. So far, this is simply a popular website that pays the costs of maintaining and upgrading content.

The *third* feature is crucial. It is the ability of the platform to *collect, retain and use data about users*, combining information from users about user behaviour with its business model. In the case of Amazon, this involves gathering information about user searches and purchases to make offers that are more likely to interest the user. The extent to which companies exercise this capability differs widely and is complicated by the existence of services such as Google's AdSense (discussed below) that allow smaller platforms to gather data (shared with Google) about their users. The full revenue-generating potential of the third element is apparent in the case of Google. Google can record information not only about the purposive behaviour of users as they search for goods and services, i.e. finding an appropriate e-commerce site, but also about all the other searches that users make (and the information they contribute) that disclose their interests and their behaviour. To the extent that Google can use this information to make better presentations of advertising (and convince advertisers of their ability to do this), it can deliver an advertising service that generates greater economic returns and competes successfully against other outlets for advertising.

Google's development of this third feature illustrates a *fourth* component in our definition of a digital platform. This is the *provision of auxiliary services* that become possible as the result of the first three features. Many of these services involve the platform serving as an intermediary in the online world. By 2003, Google was offering companies such as newspaper publishers, book publishers and other goods and services providers the ability to gain advertising revenue by allowing their websites to be used to offer advertising provided by Google's AdSense on behalf of advertisers. Google had created an auxiliary service and was functioning as an intermediary between advertisers and potentially any website owner. Performing this role allowed the company to gain an additional revenue stream, a brokerage fee charged to advertisers. Even if a user reached a website without using the Google search engine, Google still gained revenue. In placing ads, Google could use the data about users from its search engine to select advertising of greatest potential interest to the user, thereby raising the perceived quality of its advertising service for advertisers.

The term platform has become ubiquitous in the literature. Our principal focus in this book is on platforms that are most active in exploiting the third and fourth features of our working definition. This means that we emphasise the Google, Amazon, Facebook, Apple and Microsoft platforms – those with virtually unlimited ambition for growth in user participation as well as capture of data contributed by users or derived from observing their behaviour. We give less attention to companies offering platforms that conform to our working definition but in which the third or fourth feature is *incidental* to their operations. Examples of what we mean by incidental are so-called sharing platforms like Airbnb or Uber. These are platforms, but the way in which they employ data about users (supplier and end user customers or citizen users of the service) is markedly different than Google or Amazon. The most important aspect of these sharing companies is the trust that can be created between suppliers and users that results from a stream of feedback about both, prior to any transaction. Their collection, retention and use of data for other purposes is (so far) limited and, hence, incidental. A second example is Wikipedia, which is self-governed by a privacy policy that dramatically limits the use of data gathered about users and specifically commits Wikipedia not to resell this data to third parties.

To summarise, our working definition embraces definitions that are prominent in the literature within and beyond the economics discipline. As set out in Chapter 1, for our purposes a digital platform has four elements: (1) content desired by users; (2) a business model that pays the costs of maintaining and improving the platform; (3) the collection, retention and use of data about users; and (4) the provision of auxiliary services. The first three are essential and the fourth is a key element in the operation of larger platforms. As above, it is possible for the data a platform collects to be used exclusively by the platform owner without leading to the provision of auxiliary services. There are many online facilities that have the elements of our working definition, but which are likely to have self-limiting ambitions regarding the users they target, e.g. airline reservations systems and other sectoral platforms.

2.5   Examining digital platform elements

The components of our working definition are developed in greater detail in this section to demonstrate how each of the elements has specific consequences.

2.5.1   Content desired by users

In the case of the first component, creating content desired by users, a key aspect is the creation of a reinforcement cycle in which the platform operator captures and utilises user-related data arising from interactions to intensify the user experience. The aim is to encourage the user to interact further with the platform; the content is made more desirable as the platform owner learns about the user as the result of the user's interaction with the platform. This is a very powerful way of constructing large-scale online networks of users who, viewed from the platform operator's perspective, are an audience for advertising, the sale of goods or services, and other possibilities. This relatively simple idea is easier to state than to the make operational. Subsequent chapters show how, in a variety of contexts, this reinforcement cycle operates, and we trace its consequences, not only for commercial value, but also for public values.

The market-making functions of platforms and the incentives created by the platform technical architecture as well as the business models that generate revenues for the platform owner can be understood using neoclassical economic analysis. However, account must be taken of the peculiarities of data-based or information products and services. When treated as an economic commodity, information differs from other commodities. It is often expensive to produce a first copy, but the unit cost of reproduction of subsequent copies is negligible and, thus, cost is independent of its scale of use.[3] This is the principal source of economies of scale in information industries such as platforms. Costless reproduction of data-based information challenges standard neoclassical economic analysis which ordinarily is focused on the role of scarcity in the determination of price. For platforms, scarcity is created by the ownership or control of data-based information and this ownership or control is what generates the economic value of the platform. Thus, the specific societal institutions (norms, rules and standards) that govern the control and ownership of data have very major consequences for incentives to produce and distribute information that is created using data resources.

2.5.2 Business models for platforms

The foundations for these developments were established early in the development of platform business models, the second feature of our working definition. We consider two of the basic business models employed by digital platforms: user-produced content and platform-based e-commerce.

2.5.2.1 Basic business model – user-produced content

User-produced content has enormous potential for the monetisation of data which Facebook, a social media platform founded in 2004, illustrates. Facebook represents an innovative solution to the first feature of the platform – creating content desired by users. Facebook's answer was to invite users to create this content themselves. This was not the first example of a social media site where users shared information about themselves, their interests and their views. By simplifying the process of establishing a user-controlled page and streamlining the sharing of the location of this page using the idea of "friending", Facebook made it easy to invite others to see and interact with other Facebook users. Facebook's business model would become advertiser-supported so that it could offer its services freely to users. It quickly gained access to an enormous quantity of data about users – their interests, social connections and, importantly, their locations.

The opportunity that Facebook tapped into, user-created content, was a part of the larger trends discussed in Chapter 1. Any website which attracts users can develop a business model to exploit data if the owner *chooses* to do so, making it a platform by our definition. A site operating entirely with user-created data needs a business model to pay for the costs of its maintenance, including the monitoring of content that might offend users or otherwise discourage them from using the site. The business model might be voluntary donations (e.g. Wikipedia), subscription (e.g. some educational sites and newspapers), or a subsidy from government or other organisations (e.g. health, public service media). The selected business model will influence how the platform owner employs user-contributed and user interaction-generated data, including the sale of access to this data to others. As indicated above, Wikipedia chooses not to endanger its voluntary contributions by carrying advertising or selling information about its users. YouTube carries advertising and, as part of the Google family, derives revenue from using data about its users to target advertising messages to them on all the platforms with which it has a business relationship. Both platforms rely on user-contributed and user-generated content for their existence. Seen from a (neoclassical) economics perspective, however, the digital platforms, including social media, have *chosen* to use methods for data collection and capture aimed at generating economic value.

Some authors, including some of those in the institutional and critical political economy traditions, observe that platform owner use of user-created data – and even data generated by user interaction with platforms to generate value – is unjust. Without user activity there would be no data and, hence, no ability to capture value. This sort of argument may lead to the conclusion that users should receive some form of compensation from the value that is generated. A first response to this argument is that users' choice to interact with a platform is a form of consumption;

it creates value for the user. The fact that this interaction can be further developed and monetised by the platform owner is not relevant to the user's choice to interact. If it was relevant, platforms based upon different business models (e.g. donation or subscription support) would prevail in the market. The fact that these alternative business models are not prevalent may suggest that prevailing platforms offer innovations based upon creativity and investment larger than alternative business models are capable of supporting. If this is the case, then rules, e.g. collective or public management of data, that abridge the platform owners' abilities to monetise user-contributed and user interaction-generated data, present the risk of impeding platform owner innovations that make online services attractive to users. This is a risk that policy makers may be willing to take in furthering public values.

2.5.2.2   *Basic business model – platform-based e-commerce*

The combination of the virtualisation of e-commerce and the use of user monitoring (or surveillance) and user-generated content is the basis for a second powerful business model in the online environment. This is because of an economic effect, long recognised in neoclassical economic analysis: digital platforms benefit from network economies or effects. This is the principle that a network's value and desirability increases with the number of its users. In the case of e-commerce and social media platforms, buyers and sellers are attracted to larger markets than would be feasible to create in an offline environment. These network economies explain demand – the number of users wishing to participate on a platform. This is discussed further in Chapter 3 and here we focus on the costs of supplying service to these users.

Platform costs differ depending upon the function they serve. In the simplest cases, social media and online game platforms, the costs of serving users are limited to the costs of designing and improving the software defining and operating the platform, promotion costs for the platform including the costs of selling advertisements, the costs of storage for the software and user content, and the communication costs in maintaining user service. Designing and improving the software is a cost category that benefits from the economic properties of information as discussed earlier. This means that the unit or average costs of providing software, including software designed to support social media or online games, declines with the number of users. The importance of this component of our working definition is well illustrated by a comparison of physical markets and the virtual market offered by platforms such as Amazon, Alibaba and eBay.

Let's take a specific example.

The city of Turin, Italy, has 42 open-air and six covered markets. The largest of these, the Porta Palazzo market, has 800 stalls and covers an area of 50,000 square metres (over 12 acres). It claims to be the largest open-air market in Europe. This market is often crowded with buyers, but the motive for going to it is clear – the profusion of sellers offers choice and price competition. Sellers are attracted by the number of people attending the market, with the possibility of increased sales outweighing the some-what keener price competition among sellers in Porta Palazzo compared to other Turin markets.

E-commerce platforms such as Amazon and Alibaba are examples of online markets. The web-scraping company, ScrapeHero,[4] reports that Amazon offers more than three billion products in 11 markets worldwide and 564 million in the US. Let us compare Amazon's American market offering with the Porta Palazzo market in Turin. Supposing, generously, that the average number of unique products offered in Porta Palazzo per stall is 100, the market offers 80,000 products. This means that Amazon is over 7,000 times as large as Porta Palazzo. If Amazon's products were laid out in stalls like those at Porta Palazzo, its market would occupy 84,000 acres (about 34,000 hectares). The entire land area of Turin – 13,000 hec-tares – and Washington DC – 18,000 hectares – would not have enough space to display Amazon's US product offerings.

Consider what this would mean for navigating among the offerings. In Porta Palazzo there are ten different departments dividing the sale of clothing, fruit, fish and housewares. A walk of several hundred metres is enough to gather items from several departments. At the Amazon website, every product is categorised and searchable. The visitor to the Amazon platform may consider lightbulbs, then books, then clothing, without stirring from a computer. Rather than tendering money to several different vendors, the visitor may check out with a "basket" containing all the selected purchases and have them delivered within the week, or much sooner in a metropolitan area. In addition, in the event of no purchase, Amazon will have collected data about the visitor. Both revenue from sales and revenues generated from user data are used to recover the costs that Amazon incurs and, hence, are key elements in their business model. This example also illustrates an advantage of e-commerce platforms. Assuming the platform provider can arrange for the inventory of physical goods and delivery services, the visitor can locate and order products from a stunningly large collection of offerings. The shopping experience may not be as interesting or social as visiting Porta Palazzo, but it is more efficient (or convenient) in terms of the shopper's time and shoe leather.

2.5.3   Collection, retention and use of data about users

In the early development of platforms, the ability to monitor customers and retain information was confined to the customer's interaction with the platform. Later, this third feature of our working definition – a monitoring or surveillance capability – was extended and enhanced. Monitoring customer behaviour is not novel, however. Offline retailers have used closed-circuit television or in-store observers to track sampled customer behaviour. Companies developed customer relations management systems to view or analyse the history of their interactions with individual customers. The online monitoring capacities available to platform owners now are so much greater in degree that they are a distinctive feature of platforms engaging in e-commerce.

These monitoring capabilities are not confined to companies that conduct e-commerce. Social media and search engine platforms monitor the behaviour of users to target advertising, raising the economic value of advertising by making it more likely that it will be received by a receptive audience. Virtually all platforms make use of user-created content. In social media platforms such as Facebook, this user-created content is the principal reason that users sign up to and use the platform. When content (messages, photos, etc.) created by friends and family or others is viewed, the display can include advertising messages and the platform serves as an intermediary for advertisers to reach an audience.

The monitoring capabilities and the use of user-created content allows platform owners to combine their algorithms and some human intervention to select the advertising messages most likely to be relevant to a user; if a user talks about coffee on the online platform, advertisements about coffee and coffee-making equipment can be displayed to the user. This "targeting" of advertising by user interest makes the advertising placement service offered by the platform more valuable to advertisers. It allows the platform owner to charge a higher price than for less targeted advertising. Revenue from advertising is expected to pay the costs of offering users access to the platform and so access can be offered without charge.

The generation of advertising revenue is the dominant business model for most, but not all, platforms. Exceptions are platforms serving a more traditional intermediary function, like Amazon, where the advertised goods and services are also directly available for sale. The platform owner receives a share of the purchase price and has the possibility of marketing its own services to sellers. Many of these platforms, including Amazon, nonetheless, make use of user-created content in the form of product reviews. Some platforms choose to employ subscription or voluntary contribution models and may not seek to exploit users' data to the same extent as the platform leaders.

2.5.4   Provision of auxiliary services

In addition to the first three features of our working definition, the larger platforms develop and commercially exploit auxiliary services. The example in section 2.4 of Google's AdSense is a case in which these auxiliary services extend beyond the basic operation of Google's platform to include other websites that have nothing to do with search services. Although the development of auxiliary services initially was a secondary aim, it has become a major source of revenue for platform companies.[5] A prominent example is cloud-based services. Since platforms exist as information architectures on web servers, it was an obvious business decision to operate those servers or server farms rather than to contract with external parties for these services. At Google and Amazon, the services include website hosting, cloud-based enterprise software and configurable computational capacity.[6]

By a slightly different route, the development of enterprise software, Microsoft also offers a substantial range of cloud-based services. Facebook's and Apple's auxiliary services are more closely linked to their consumer-oriented services. While some 85 per cent of Facebook's revenues come from advertising, the company also receives revenues from games and sales of data analytic services to other companies. Apple's iCloud is a file storage service like those offered by other platform companies, e.g. Google Drive. After a modest initial allocation, users can pay a subscription charge to increase the amount of online storage. All these companies offer recreational and other software, with Apple, Microsoft and Amazon selling or offering software with a focus on the devices these companies supply, while Google and Facebook also sell software. The auxiliary services mentioned here are also offered by other companies that do not necessarily function as platforms. Perhaps the most important area of future development for platform companies is the delivery of AI-based solutions. This is examined in Chapter 4.

2.6   Digital platforms, norms and values

Developments which made the digital platform a reality were shaped by private and public decisions concerning the norms, rules and standards by which the providers of online services can operate (we discuss the economic analysis of these in greater detail in Chapter 3). It is important here to acknowledge the role of these norms, rules and standards, some introduced by governments and some the result of combinations of government and corporate decisions.

During the early development of online services from 1995 when all commercial restrictions on the internet's use ended, and the financial crash in 2008, government policy was largely welcoming of the digital platforms. This was because it was believed that new online services offered potential for economic growth and employment in an "information" or "digital" economy.[7] Due to a fear of being left behind as new opportunities developed, government-initiated rulemakings sought to reduce apparent barriers or obstacles to the growth of these services. For example, e-commerce relies on systems for customers to purchase goods online or for suppliers to contract with each other or platform providers. Governments enacted legislation assuring that such payment and contract methods would be comparable to their offline equivalents in terms of enforceability and liability.

Governments also responded to early digital platform entrants' concerns about their responsibility for user-produced content that might be libellous or otherwise problematic. Western governments typically interpreted the role of platforms as a neutral intermediary and provided the platforms with a "safe haven" from liability. This was later abridged by legislation requiring platform owners to remove (take down) content infringing copyright or otherwise found to be problematic once they became aware of it (we discuss this in Chapter 6).

When the norms and rules that influence public values and market outcomes are considered, there are many reasons to be less sanguine than government policy makers were in this early period. The growth of the digital platforms began to disrupt existing businesses that were disadvantaged by being tied to physical locations. For example, brick and mortar stores employing salesclerks and workers to maintain physical inventory began to find it difficult to compete with large warehouses maintained by platform providers capable of supplying goods directly to customers. The "place-less" nature of digital platforms started to raise issues about taxation (sales or value added tax as well as other forms of taxation such as property or business taxation). Proposals were made to enact a "bit tax" (Soete and Kamp, 1996), a means of mitigating some of these disruptive effects. These were rejected on the argument that this would impede innovative growth of a new type of business which might prove to be more efficient and provide opportunities for entrepreneurial business and for beneficial social and political communication.

Other features of digital platforms that would prove problematic for security and privacy were inherited from the internet. The technical standards for security and privacy prevailed because more rigorous methods were treated as a potential barrier to the internet's universality and growth and as inconsistent with the desirability of an open network. This meant that privacy and security had to be dealt with by internet users. Business users (and their suppliers) were reasonably well-prepared for doing this. They had experience with electronic trading networks that were often implemented using proprietary communication protocols offering privacy and security features. These could be reproduced using various forms of encryption on the "open" internet among users who agreed to such standards. Some limited security was also achieved for less technically sophisticated users in areas like electronic payment systems. Individual privacy, security and safety issues continued to be raised as user surveillance became a principal business model for enhancing the value of advertising during platform growth.

The emergence of e-commerce and the norms and rules informing the digital platforms' operations were of concern, additionally, because of the potential for these platforms to disrupt labour relations. A long history of struggle during the industrial era had led to better (but still often deficient) wage settlements and working conditions for industrial workers. These gains had not been achieved by workers in the service sector for reasons including the rapidly changing industrial structure of wholesale and retail trade and the diversity of other service industries. Resistance to union organisation may also be attributed to class distinctions between blue- and white/pink-collar workers. Labour solidarity was less broadly shared among the latter group, which resisted classification as working class (Frey, 2019). Yet in the 20th century, the service sector had become the principal source of employment in the global North. Claims that the platforms' norms and rules were generating more "efficient" ways of responding to market demands were criticised as value-laden claims. These criticisms provoked questions about the distributional consequences of "improvements" in efficiency, especially when the platforms' operations could be shown to be associated with a worsening in the pattern of income distribution. This served the interests of capital owners well, but immiserated or marginalised workers. Thus, the lowering of standards for the treatment of workers and projections for the growth of precarious forms of employment are being amplified by further development of digital platforms.

User-produced content also raised issues around the monetisation of these user contributions. The platform owners were receiving value from the "voluntary" contributions of their users. They were, in fact, benefiting from "unpaid labour", a further shift in income distribution towards the corporate sponsors of platforms. Critical analysis focused on how these developments represented another facet of the exploitation of the many for the benefit of the few. As it became apparent that not only user-contributed content but also data resulting from monitoring user search and other online behaviours was being monetised, some interpreted this as extending the scope for the exploitation of individuals within Western capitalist markets. The seeming "placelessness" of digital platforms (Flecker, 2016), permitting the circumvention of national taxation and, in some cases, national government legislation, rather than being treated as a defect in governance norms and rules subject to certain reforms, was treated as an instance of the consequences of global financial capitalism and the "hollowing out" of state authority except with regard to security issues (Gurumurthy et al., 2019). All these concerns were also reinforced and amplified by platform growth.

2.7   Conclusion

This chapter has discussed digital platforms as radical innovations. We have highlighted specific developments that gave rise to today's platforms and provided a working definition that informs the rest of the chapters in this book. Our account of the origins and novelty of platforms makes it clear that a very broad scope of human activities is facilitated by digital platforms. This means that an understanding of the consequences of platform developments needs to extend beyond the scope of neoclassical economic analysis, e.g. its usual neglect of social consequences such as income distribution or the quality of work. This neglect occurs because such issues are taken to be matters of political choice or are outside the frame of analysis. However, the scale and scope of digital platforms in society requires an analysis of their operations that is additionally concerned with the choices and consequences of platform behaviour. In the next chapter we discuss insights that neoclassical economics provides and how two additional strands of economic analysis are essential additions for understanding platform economics.

Notes

1. A website is controlled by virtue of the web server's protection of files from alteration. This protection is not entirely reliable because it depends upon the security provisions of the web server and because the content of the website can be duplicated and located on a different web server controlled by someone other than the website's author with possibilities for good, e.g. the Way Back Machine (an archive of web pages) and ill, e.g. schemes for committing fraud. If users can be misdirected to the non-authentic site, they are at risk of being deceived.
2. There are some parallels with telephone directories, which offer ordinary and, for a fee, enhanced listings.
3. Unit costs or average costs are total costs divided by the number of units. The first copy cost is incurred only once; subsequent copies are nearly costless; the more copies produced, the lower the unit cost, see Arrow (1984).
4. Web-scraping is the practice of using automated means to extract data from websites. See ScrapeHero (2019) and Black (2012) for a scholarly study of the market.
5. In 2018, Alphabet's revenues from AdSense and its newer MobAd (mobile service advertising) were USD 19.98 billion or 14.7 per cent of revenues earned from its main sites. In the same year, other revenue from user services

and cloud-based service provision was USD 19.91 billion (14.6 per cent) (Alphabet, 2018: 25).
6. In 2018, Amazon Web Services received USD 25.66 billion or 12.4 per cent of its domestic and international revenues from other operations (Amazon, 2018: 23).
7. See Bauer and Latzer (2016), Brynjolfsson and Kahin (2002), Goldfarb et al. (2015), Mansell (2010) and Mansell et al. (2007) on information or digital "society" and "economy" developments.

3. Economic analysis of platforms

3.1 Introduction

Digital platform operations are a part of larger processes referred to as digitalisation and datafication, as we have seen in the first two chapters. We introduced a working definition of digital platforms and explained their origins. These involved the translation and adaptation of traditional processes into the online world and the invention of new processes and relationships. This chapter examines how three strands of economic analysis – neoclassical, institutional and critical political economy – have been adapted or developed to explain and interpret the digital platform phenomenon and its basic elements.[1] Key insights from this chapter are the tensions between market-oriented logic and public values which are heightened by datafication; the need to question the boundary between private and public interests; and the role of power in shaping processes and outcomes. In Chapter 4, we will see that it is not only these basic elements, but also opportunities afforded by technical and business innovations related to AI, that help to explain the continuing growth and influence of digital platforms and their consequences for public values.

3.2 Neoclassical economic analysis

The explanation of digitalisation and datafication in neoclassical economic analysis starts from the view that the exogenous arrival of a new technology is a potential "engine of growth" (Bresnahan and Trajtenberg, 1995). The origins of major or radical innovations are not explained by the neoclassical analytical framework although there have been attempts to make investment decisions in new technologies part of the framework of neoclassical growth models (Romer, 1990). Most commonly, the framework assumes that such innovations draw upon the knowledge produced by public and private investment in the general advance of knowledge. This investment sometimes yields unanticipated innovations of a radical nature. Digital technologies are expected to have large growth effects because they create new goods and services as well as offering potential efficiency gains in the production and distribution of existing ones.

Neoclassical economic analysis employs key assumptions that are useful for the application of logic to derive clear conclusions. A central assumption is that human behaviour is governed by the pursuit of the goals of profit and individual satisfaction.[2] Every individual is assumed to have an endowment of capabilities and preferences that is generally taken to be given (outside the frame of analysis). Because society contains many individuals, there is a very large number of possibilities of exchange for furthering these goals. In its most simplified form, neoclassical economics assumes that individuals and companies are aware of all possibilities for exchange. Individuals choose to supply labour to the employer (or activity) which will lead to the best satisfaction and wage (or income) given her or his endowment of capabilities. With the money earned by supplying labour, the individual chooses among available goods and services those that yield the highest satisfaction (which is called rational choice). Exchange is idealised by assuming that a market exists in which suppliers and customers make exchanges based solely on price and qualities of the good or service.

Markets are assumed to be competitive, which means that there will be a prevailing or market price of individual goods and services. The prevailing price is determined by the total demand of all customers and by the aggregate willingness to supply of all suppliers at the prevailing price (the familiar "scissors" of demand and supply). The existence of a prevailing price is a direct implication of the pursuit of profit. A supplier deviating from the prevailing price by charging more will face competition from other suppliers seeking (the ever smaller) profit to be gained by winning customers away from the deviating supplier.[3] The prevailing price is ultimately related to the cost of production, which includes a "normal profit" – the amount of profit (adjusted for risk) that will encourage investors to provide the supplier with the means of production (including the fixed capital of premises and machinery and the working capital to acquire labour services).

The preceding two paragraphs are an outline of what is sometimes referred to as "textbook" economics, the basic neoclassical framework. Professional economists routinely vary these assumptions in order to analyse features of the world that do not conform, or conform poorly, to the textbook assumptions.[4] In this book, the most important variance of assumption concerns information. In reality, suppliers and customers are not completely aware of all exchange possibilities. Advertising provides information about exchanges that could be made and, somewhat more controversially, that may influence the preferences of customers. Allowing information to have economic value by assuming it is unevenly distributed opens the door to considerable complexity as well as difficulties in observation.[5] For example, choices cannot be presumed to reflect true preferences if individuals are unaware of alternatives and may reflect biases or beliefs about suppliers. Even more troubling, if information can influence preferences, it is possible for a particular supplier to raise the price for their good or service so long as competitors are unable to exercise similar influence. The uneven distribution of information is one of several possible sources of "market failure" – markets failing to perform according to the assumption that they are neutral places of exchange between suppliers and customers. A more prosaic type of market failure is the possibility of supplier collusion, which generally is assumed to break down because other less cooperative suppliers will enter the market to overcome the collusive firms.

The following sections (3.2.1–3.2.5) outline some of the key concepts employed by all three approaches to economic analysis that we consider in this book, concluding (3.2.6) with a brief summary of the limitations of neoclassical economics. This is followed by sections outlining how the two other approaches – institutional economics and critical political economy – vary the basic assumptions in the interests of capturing features of economic life that can be oversimplified by "textbook" economics or by professional economists who generally retain key basic assumptions.

3.2.1   Economics of multisided markets

Digital platforms allow the platform owner to offer a multiplicity of services, as in the case of the provision of social media services. Users post content (user-produced content) and platform owners recombine this content with advertising (advertising content) for which advertisers pay the platform owner. The revenue from advertising may allow the platform owner to offer the posting service to users for free (as explained in Chapter 2). In this case, there are two markets – one for postings offered to users and another for advertisers, offering them a channel for their advertising messages to customers. This is treated as a two or multisided market in neoclassical economic analysis to indicate that the platform is operating as an intermediary or matchmaker service for two separate groups (e.g. users and advertisers).

Multisided markets complicate the usual neoclassical economics assumption of an atomistic competitive market with a multiplicity of sellers and customers. The complication is that neoclassical economic models of supply and demand generally do not account for the interdependence of each group which is making use of a platform on its different sides (Evans and Schmalensee, 2016). Instead of markets being assumed to stand apart from the operation of a digital platform, the platform operator can shape market relationships that occur on the platform. Individuals are assumed to choose to participate on platforms because the intermediary platform provider reduces frictions that make it more costly for participants in the market to interact directly (e.g. by e-mailing their content to their friends). The benefits of participating in the exchange of such content (self-publishing and viewing content produced by others) may, if well managed, increase with the number of users and, thus, the digital platform innovation may support a self-reinforcing feedback loop.

The consequences of this feedback loop are called network economies or network effects and result in the growth of value (to both platform owners and customers) with increases in the number of users.[6] In this process, the user is assumed to tolerate, or even to value, the addition of advertising content if the postings and display of content services are efficiently organised. While users generally contribute content without compensation, in recent practice, "celebrity" users contributing content attracting many viewers may be offered a share of the advertising revenue this generates. In other words, some users may become content suppliers comparable to authors of games or other diversions that accompany the user-created content as attractions for the platform. Thus, central to the multisided digital platform is its role in sharing data to support its customer services. Efficient display of content means that its quality or meaning is filtered in various ways, including allowing users to post links to existing content.

Data-intensive digital platforms (such as those based upon user-created content) scale up by attracting participants on both sides of their platform (more contributors and more advertisers). Because of network economies, scaling up is expected to be non-linear, e.g. for an x percentage increase in users, the percentage increase in value and desirability of the platform is y, where y is greater than x. These economies are reinforced by *economies of scale* in supply, that is, when unit costs decline at a large scale of output because of the economics of information noted earlier. In addition to economies of scale, platforms also benefit from *economies of scope* in demand or supply – the property of adding additional products or services leading to greater total demand or a reduction in average cost of each of them, compared to offering (or producing) each separately.

The generation of economic value in the multisided market requires datafication. Due to network effects and economies of scale and scope, platforms can achieve a level of user participation that consolidates their position in the market, further increasing their capacities for datafication. The rate of increase in a network's economic value and desirability from the platform owner's and, it is assumed the users' view, depends upon the nature of the platform. Some platforms may experience stronger network effects than others. When the network effects are large, it becomes difficult for others to compete without the scale and scope effects resulting from the dominant platform's large user base and control of data. A platform's dominant position in the market reduces the incentive for users to switch to a different platform (even when they can sign up to multiple platforms) unless the services of a competitor are substantially different. The interpretation of this dominance is contested. Some economists argue that the strength of network effects leading to dominance is the consequence of innovative ingenuity in the design and operation of the platform (Bork and Sidak, 2012). Insight into the argument about dominance is provided by considering the role of economies of scale that accompany the scaling-up of platform operations.

3.2.2   Economies of scale consequences

Scale economies affect a platform's cost structure and its capacity to build a dominant presence in the market. When a platform hosts content (either user-produced or produced by others) that attracts users, the larger the audience, the smaller the per person cost (average cost) of storing and delivering this content. This decline in average cost will also apply to the costs of software design employed by the platform owner. So far, we are only considering the peculiar economic features of data – their expansibility or reproduction at negligible cost – to indicate the existence of economies of scale (as discussed in Chapter 2). However, it might seem that data storage costs would increase with the amount of content or users, or both. This might be a barrier in the extreme example in which every piece of content was accessed by only one user, similar to a fantasy of a library that receives books but which no one ever visits.[7] However, because much of the content will be accessed by multiple users, the economies of scale provided by the expansibility of information will dominate the diseconomies (additional costs) of additional storage. In any case, even a large collection of instances of small audience content may be sent the same advertising message.

An interesting feature of platforms, dubbed "the long tail", is that the number of user accesses to content will follow a declining distribution with some content receiving very large numbers of accesses, followed by a tail in which access declines (Anderson, 2006). Even though receiving fewer accesses, the content located in the long tail contributes to the value of the platform for users. Maintaining the long tail is enabled by technological advance which is driving down the costs per unit of data storage at a rapid rate. Thus, even though the addition of content imposes additional storage costs, these costs are falling over time. Finally, the costs of communicating data are also subject to economies of scale and to additional economies related to the economics of the underlying communication network. For example, the growing use of fibre in the telecommunication infrastructure is reducing the cost per unit of transmitting data.

Digital platforms engaged in e-commerce incur additional costs related to maintaining an inventory of products and the delivery of products and services. The physical space required for maintaining an inventory of products would be very large if they were displayed to customers, as illustrated in Chapter 2 by the comparison of Turin's Porto Palazzo market with Amazon. Instead, what is displayed is an advertisement for the product containing its image, descriptive information and, perhaps, user reviews. The goods can be densely packed into a large warehouse organised to optimise the finding and shipment of customer selections. There are economies of scale in such warehouses and in establishing networks for the delivery of products and services. These economies are not as marked as those benefiting social media or other digital content platforms, but they are sufficient to make platforms like Amazon strong competitors against offline retailers of products and services.

3.2.3   Natural monopoly claims and counterclaims

Strong increases in user interest as the size of a digital platform grows (i.e. strong network effects), accompanied by economies of scale and scope, define a type of market structure – the *natural monopoly*. A natural monopoly is said to exist when economies of scale continue indefinitely so that the largest firm operates at lower average costs than any actual or potential rival. The "naturalness" of this monopoly is a feature of an unusual technology, one that offers indefinite cost reduction as scale increases. In the case of a social media platform, if visitors sought to do only one thing, e.g. communicate with other users of a platform in the form defined by a Facebook-like interface, then network economies and the economies of information or data storage and transmission suggest that Facebook is a natural monopoly. It can serve an ever-increasing number of users at average (unit) costs lower than any smaller rival.

This logic often leads to the conclusion that some digital platforms may be a natural monopoly in the provision of applications (e.g. social media or e-commerce). The counterclaim is that forces of innovation-enabled competition make this position vulnerable and potentially ephemeral. There are several possible arguments for this counterclaim.

The growth in the size of a digital platform might suggest a "diseconomy" (a higher cost accompanying larger size). This is the cost of search. Platforms do reduce search costs. A customer might have to examine many stalls at an offline market to find a product. Amazon reduces these search costs by a detailed classification of product types, but users may still experience some diseconomies. A person looking for a light bulb must be careful to match all the desired characteristics with the many products on offer. Offsetting this is Amazon's ability to offer specialised light bulbs that smaller suppliers may not be able to supply. Amazon's product categorisation and its provision of technical specifications for each product is one way of reducing search costs. However, it cannot be ruled out that a competitive e-commerce platform might offer alternatives. An example is music and video streaming, both of which are offered by Amazon, but many competing sites have emerged to offer similar services (e.g. Netflix, Hulu, Spotify, Deezer, Apple Music, Google Player, Vevo and Tidal). The extent to which these alternatives operate as platforms (e.g. observation of user behaviour to target advertising) may differ, but their survival indicates that enough users exist to sustain these services, each of which has somewhat different designs of the user interface.

There are also trade-offs between a large audience, some of whom will be interested in an advertising message, and a more specialised audience, containing more individuals with an interest in a specific advertiser's message. Platform owners attempt to compensate for this by developing predictive models about their users so that they can stratify or identify those most interested in the advertising messages of each advertiser (discussed further in Chapter 4). It is possible that this strategy will be less efficient for some types of audiences than a strategy aimed at creating a website that will attract people with specific interests who also are a more likely audience for advertising content. The strategy of attracting a specialised audience has long been employed by magazines where, for example, the content of fashion magazines is stratified according to the age and income (real or aspirational) of their readers. For those who are sceptical of natural monopoly claims, this possibility for entry of competing platforms is evidenced by the numerous social media platforms that exist despite the purported scale advantages of Facebook.

Digital platforms are also vulnerable to malicious and opportunistic behaviour. The larger and more prominent the platform, the more difficult it is to govern this behaviour and the more attractive the platform may be to individuals or groups seeking to do harm. Such behaviour occurs when individuals seek to gain, e.g. by manipulating customer reviews on e-commerce sites, or simply to engage in malicious behaviour, e.g. posting vituperative and libellous claims on social media platforms. The platform owners have sought algorithmic solutions to mitigate these behaviours. However, this provokes an "arms race" with the opportunistic and malicious actors inventing workarounds such as registering for multiple accounts or finding other points of vulnerability in the platform (Taddeo and Floridi, 2018). The costs of governing platforms to prevent malicious and opportunistic behaviours are a possible diseconomy of scale, and, the larger the platform, the higher the costs of mitigation measures are likely to be.[8] This is so particularly when these measures involve hiring human moderators to vet user-produced content for violation of terms of service agreements or codes of practice.

Because of the robustness of competition, neoclassical economic analysis concludes that, while natural monopoly is possible, it is also possible that areas that might be "monopolised" can be contested by innovative strategies. Whether market dominance *will* be contested is an empirical question. The answer can only emerge from market trial and experience. Most individuals wish to do many different things online and some prefer not to use, for example, Facebook, to communicate with others about their lives. Even if Facebook experiences continuing unit cost reductions, limits to its growth are present. These limits so far are not very constraining since Facebook had some 2.4 billion monthly active users in June 2019, about 30 per cent of the world's population, and was larger than any other online site.[9] However, even though there are other smaller platforms in the market that have yet to be acquired by Facebook or other dominant digital platforms, the largest platforms are disrupting on- and offline businesses with which they compete for advertising revenue. Next, we consider some of these consequences for the media industry and for e-commerce.

3.2.4   The media industry

The scale and scope economies experienced by digital platforms are destabilising the traditional media industry (e.g. newspapers, television, radio and magazines). Over the past half-century, the introduction of new entrants has meant the share of advertising spending on pre-existing media has declined. This was notable with the introduction of television, which reduced the share of total advertising spend on radio, newspapers and magazines. Internet advertising, which is mostly advertising on platforms, has dramatically affected the distribution of advertising spending. In 2018, internet advertising absorbed 49 per cent of all advertising on all media in the US and 45 per cent in Europe.[10] An unresolved issue is what this does to the quality, plurality and diversity of media content and, especially, news content. The financing of newspapers differs substantially across countries, but subscriber and newsstand purchases of newspapers amounted to only a third of newspaper revenue in the US in

2017. Similarly, radio and television journalism principally are supported by advertising, except in countries with a commitment to public service media (or state-owned media). The result is a weakening of the revenue base for certain genres of digital content, including traditional investigative and other forms of journalism (addressed further in Chapter 4).

3.2.5   E-commerce

The competitive pressures introduced by digital platforms have significant implications for the structure of employment in the retail and related wholesale trade sectors. The computer coding and systems maintenance jobs required by digital platforms are lower as a share of a platform's revenue than the labour inputs required to maintain physical operations in virtually any other type of business. E-commerce platform operations involve stocking, picking and delivering products from large warehouses. In these warehouses, workers can be employed intensively in highly routinised activities assembling and packing orders, often aided by automated or mechanised equipment including robots. In the near future, these workers may be replaced by robotic systems (the implications of which are discussed further in Chapter 4).

These competitive pressures have implications for jobs. Smaller numbers of retail salesclerks are required as the volume of "brick" outlets falls relative to "click" outlets. This is relatively gradual because of features of physical shopping that are not easily matched online. These include the social experience of shopping (including interaction with human retail sales personnel), the physical inspection of goods prior to purchase and the immediacy of the purchase/ownership experience. It is also gradual because it does not follow that, because a platform can operate at lower cost, it will price goods more cheaply than a brick retailer. The knowledge that platform owners have about customers as a result of their data analytics may be used to set higher prices or guide users to more expensive products.

3.2.6   Limitations of neoclassical economics analysis

The market effects of platforms in a neoclassical economic analysis can be explained in terms of economic competition when natural monopoly does not prevail. The consumer is assumed to be sovereign, i.e. customers are, or should be, free to choose what they purchase, and from whom they purchase. Digital platforms are treated as neutral gateways between consumers and multiple applications, with consumers gaining benefits from information access, and, potentially, additional value from advertising accompanying this information. The platform serves to stimulate investment and create incentives for entrepreneurs to add complementary products and devise additional auxiliary services, and the growth of a platform is seen as the consequence of innovation. What is defined as "desirable", by whom, and with what consequences, is complicated when our interest is in what neoclassical economics treats as "externalities". As Tirole (2017: 57) says, on questions of what kind of society is desirable "an economist has little to say, except as an ordinary citizen".

Externalities are crucial, however, because digital platforms do not operate in a bubble of supply and demand that is destabilised by a radical innovation. A platform's use of digital technologies and its reliance on user data could either enhance or detract from public values that are not simply the aggregation of individual preferences. Considerations of inequality and justice or common societal interests are some of the externalities that are not part of the neoclassical analysis, or such considerations are skewed heavily towards the private interests of market suppliers. It *is* recognised, however, that platforms may arise out of forms of collective action rather than from private companies, e.g. communities of volunteers or publicly funded groups (as discussed in Chapter 5), and these may compete for user attention and participation with commercial platforms. Those who own platforms can set terms for platform access and use, and for the owner's use of the data that a platform generates. *In theory*, they cannot overreach in ways that will set off a mass exodus, but once a dominant position is secured, they can treat individual agents (advertisers or users) arbitrarily.

When the analytical focus is on technological innovation as the "driver" of change, the broader conditions that give rise to power asymmetries are neglected. Therefore, an understanding of platforms as a radical innovation requires a deeper understanding of what gives rise to power asymmetries and why platforms favour business models that foster the collection and analysis of data on user online choices. The workings of asymmetrical power are largely excluded in the neoclassical analysis because of the foundational assumption that markets are competitive. In this view, disproportionate power can only be exercised by an individual or company when there is a breakdown in competition or market failure, as in the case of natural monopoly or collusive or other forms of anti-competitive behaviour. The unit of analysis is the individual, consistent with methodological individualism. Public values are understood as an aggregation of individual choices because some externalities are ignored. The best means to exploit opportunities afforded by datafication is private ownership so that data can be "mined" for its economic value. The platform owner is assumed to have the right to establish norms and rules to maximise economic value. Accountability rests with the private owner of data and information derived from it because the primary interest is to advance the value of the platform to users, thereby prevailing over competitive alternatives. The role for the state is to reform any norms or rules that create friction in the marketplace. Individual agency allows consumers or citizens to interact with the platforms in whatever ways they find to maximise the benefit to them.

The foregoing assumptions of neoclassical economics concerning the ability of competition to limit power are not foundational for the other approaches to economic analysis that we consider in this book. Both employ more nuanced understandings of power and its consequences. Although the arrival of a radical innovation in workable form is largely unanticipated, and some of its consequences can be analysed using a neo-classical economic lens, these innovations involve substantial restructuring, new norms and rules, and new social, political and economic relationships. As we have indicated, these transformations lie largely outside the neoclassical lens. They are central features in institutional economic and critical political economy analysis. We start with institutional economic analysis.

3.3   Institutional economic analysis

Institutional economic analysis is concerned with the market as an institution comprised of norms and rules (Hodgson, 1989; North, 1977; Rutherford, 1994; Spithoven, 2019). Institutions are treated as systems of socially embedded norms and rules involving shared expectations. When these are not widely shared, various forms of disruption are likely. Such norms and rules typically are assumed to be relatively durable when they are consistent with a prevailing distribution of power in society. They are subject to change when this is not the case. When they do change, the change is not necessarily comprehensive or coherent since norms and rules are layered – they differ across jurisdictions, organisations, and even groups within society and between societies (Groenewegen et al., 2010). Changes which occur in one layer or one jurisdiction often create dissonance or destabilisation in other layers or jurisdictions that provide starting points for further change. Institutional economic analysis accords a larger role for public choice and public values in such changes than the neoclassical perspective and, typically, it is more concerned about the boundary between public and private responsibilities. A role is granted for private ownership, subject to arguments that may be made for a consideration of alternatives, particularly if private ownership is shown to disadvantage certain individuals or groups. In addition to setting the norms and rules to minimise market frictions, the state is understood to have a role in protecting rights and holding private enterprises to account, while also ensuring that market conditions favour prosperity and innovation.

Institutional economic analysis is concerned with institutional design, processes and practices (Williamson, 1975, 2000). The aim is often to use empirical insight to redress inequalities and seek fairer outcomes as compared to those following from existing power relations (Winseck, 2016). The question is how should market operators be held accountable for social, political or economic outcomes?

The starting point for the analysis of digital platforms in the tradition of *new* institutional economics is like neoclassical economics insofar as analysis focuses on the competitive behaviour of companies in the marketplace. The strong assumptions of conventional neoclassical economic theory about homogeneous individual or agent preferences are abandoned so as to acknowledge that "ideas, ideologies, myths, dogmas, and prejudices matter" (all of which are properties of groups as well as individuals and whose effects are therefore amplified by their collective nature) (North, 1990: 362). With a still-narrow focus on the competitive marketplace, proposals may be made for enacting new norms or rules that might avoid or weaken the "winner takes all" or "natural monopoly" character of digital platform competition. The aim might be to create better conditions for market entry in the face of a dominant platform without suppressing innovation or giving rival digital platforms an unfair competitive advantage. For example, the impact of digital platforms on the creative industries might be examined by considering the norms and rules that govern decisions about whether to rely on the commercial market as the most efficient way of coordinating and distributing online services. This determination is made by focusing on the costs of transacting among multiple digital content producers. The starting point for analysis – as in the neoclassical tradition – is the shock created by a radical innovation and how this affects producers and consumers as buyers.

In other traditions in institutional economic analysis questions are asked about how digital transformation occurs and why it matters. Questions are raised not only about economic competitiveness but also regarding the norms and rules that underpin both private and public values in society and the changing power relations among participants in society. Norms and rules are treated as being embedded in institutions (markets and other organisations) in societies. Institutional analysis grapples with a complex world of cultural, social, political and economic expectations, examining how they bear on the creation and distribution of economic value and on public values. The competitive market ideal imagined by neoclassical economists is treated in this analytical tradition as the exception rather than as the prevailing situation in economic life. Institutional economic analysis, because it draws on a rich array of social science disciplines, attends to the materiality of digital platforms – their affordances (what they make it possible to do) and how these are implicated in societal changes and in the meaning of data or information for individuals and groups of individuals. Thus, asymmetrical power relations which influence the development of digital platforms are treated as arising from multiple sources, not simply from the "shock" of a new technology. The interests of individual and groups are investigated to determine how digital platforms, as a radical innovation, are associated with positive and negative outcomes for different groups and individuals.

A central question is whether digital platform markets are leading to consumer or citizen harms and misalignments of platform operations with public values. If confirmed, institutional economic analysis is likely to assess the potential for changes in norms and rules, including the possibility that certain activities cannot, or should not, be monetised by the platform operator. Proposed intervention in a market established in response to prevailing norms and rules that, in practice, creates privileged market positions requires a thorough empirical analysis of intended and unintended outcomes (Bauer, 2014). This analytical tradition, because of its understanding of norms and rules, acknowledges that externally devised rules and norms are always present in some form and have a fundamental role in shaping behaviours and expectations.

When the platforms are found to be engaged in behaviours that limit or prevent competition, thereby creating and maintaining their privileged positions through the overt exercise of power or through covert coercion, a policy or regulatory response can be justified. Consider, for example, the case of a platform supplier being required to make an exclusive contract with the platform. This may be innocuous behaviour in a neoclassical economic analysis because of the assumption that other platforms will be willing to purchase the supplier's services. If other platforms do not exist, the willingness of suppliers to offer their services to other potential platforms may encourage entry by new platforms. It may also be argued that the potential for entry is enough to prevent a platform operator from excluding suppliers that would offer valuable services. From an institutional viewpoint, it is an empirical question whether such entry is likely.

In an institutional economic analysis of platforms, the focus on the potential for the dominant digital platforms to control gateways to digital content and other kinds of data calls for the analysis of asymmetrical power relations that enable the platforms to organise content and unduly influence people's information environments. This analysis intersects with legal and other disciplinary studies of how this power may present itself in the form of filter bubbles or echo chambers that can distort opportunities for public debate (Pariser, 2011; Sunstein, 2009).[11] Analysis of power asymmetries provides a basis for identifying how digital platform use of data-driven business strategies and operations enables them to destabilise commonly accepted norms and rules which have aimed to preserve fundamental human rights. Their capacity to operate through non-transparent collection and processing of data is a powerful constructor of reality, making alternative approaches to datafication seem distant and imaginary. This analytical lens leads to the examination of datafication and its implications, acknowledging that people may appear to derive pleasure or other rewards from their online experiences. Biases and distortions created by power imbalances, however, mean that digital platforms are implicated in adverse societal changes that should be addressed by revising or strengthening norms and rules governing platform operation.

3.4   Critical political economy analysis

In a critical political economy analysis of digital platforms, it is assumed that the varieties of contemporary capitalism share an exploitative class-based character based on the asymmetric distribution of power between those who control capital and the rest of society. This tradition locates the principal explanations for the origins and practices of platforms in the history of capitalism (McGuigan and Manzerolle, 2014). This history yields norms and rules that replicate or increase inequalities in society and analysis focuses on how struggles over platform ownership and control create "platform capitalism" (Srnicek, 2017). Empirical work focuses on how digital platforms are differentially experienced by dominant and exploited classes and how specific structures and processes of oppression are informed by power asymmetries, assuming that all operations of the capitalist market embody exploitative power relations that manifest in coercive or violent ways (Couldry and Mejias, 2019; Ricaurte, 2019). The predominance of private ownership, advertiser-based business models, exploitative labour processes and opaque algorithms is understood to limit, or exclude, opportunities to operate the commercial datafication process in a non-exploitative way.

Under this analytical lens, users of digital platforms create economic value for the platform owners by producing content that is sold as a commodity to advertisers. Users (consumers and citizens) constitute an audience that labours on behalf of advertisers and platform owners, producing economic exchange value and surplus value, and there are differing accounts of how this mechanism works (Fuchs, 2015; Hesmondhalgh, 2019a). The process of commodifying data involves platform operators in influencing beliefs and action, consistent with their interests in profit and departing substantially from neoclassical assumptions that beliefs (and preferences) are independent of the economic system. This is illustrated by the intense efforts to combine algorithms and data to enhance market "efficiency" in ways that exploit individual platform users and classes of users (McGuigan, 2019). The platform aim is to make online participants visible to measurement instruments so that data can be bought, sold or otherwise acted upon by the platforms and/or by states. For these reasons, platform owners constantly seek novel ways to extract surplus value from the activities of those who can be attracted to their platforms.

The platform owners (or managers) are assumed to exploit markets in which they reap disproportionate gains at the expense of non-owners using a host of subordination tactics, augmented by their control of data and their management of worker relations. The coercive power of financial capital is also implicated through norms and rules favouring the platform owner (consistent with neoliberalism). In some applications of critical political economy, the dynamics of datafication and platform operation within capitalism are recognised as the result of very contradictory norms, rules and practices (Freedman, 2015). Neoclassical economic analysis of commercial datafication treats the power of the platforms as an opportunity to serve customers better. In a critical political economy analysis, this power is seen to be the result of a commodification process that takes precedence over public values and negates citizens' rights and weakens their capacity to act. Platform users inadvertently contribute economic value by generating a stream of observable data on their behaviours and choices. The contradictions or dialectic of structure and individual/collective agency in the platform-mediated environment can give rise to resistance to the material and ideological determinants of exploitation. In this case, investigation centres on whether resistance can yield alternatives to platform-inspired capitalism that would lead to emancipation from the exploitative practices of the digital platforms. For instance, if it is not assumed that data resulting from platform surveillance is a resource over which corporate ownership rights inevitably must be asserted, data activism and data justice movements can be expected to develop new norms and rules for platform accountability and to restore both individual and collective autonomy (Ananny and Crawford, 2018; Hintz et al., 2019).

3.5   Conclusion

Each of these traditions in economic analysis – neoclassical, institutional (older and newer) and critical political economy – involve working assumptions about the primacy of economic and public values in policy and regulatory processes. The predominance of neoclassical economics means that the default assumption is that competitive markets, operating through decentralised individualised decisions, are best placed to achieve efficiency and an optimal allocation of society's resources in a way that is responsive to all interests and values. Evidence of market failure is the key determinant of whether changes in the norms and rules governing the platforms' markets are needed. This is consistent with the neoclassical emphasis on economic individualism. Even when public values are considered in this framing, it is the "private value of public things" that is assumed to constitute the public interest (Bozeman, 2002: 164).

The other two economic traditions are likely to acknowledge that the public interest involves collective interests that may be understood as plural (institutional economics) or as fractured by class-inflected interests (critical political economy). The issue is how these interests can be taken into account by policy makers. In the case of responses to the platforms' activities, as we discuss in the final chapter (Chapter 8), it is necessary to take a normative position on the way contested private and public values are or should be prioritised. This means that procedural norms, rules and standards are essential to establish what principles and criteria should be used in establishing arrangements for digital platform markets, and this involves choices (Cammaerts and Mansell, 2020). It is for this reason that it cannot be assumed that there is a singular public interest position that will protect individual and collective interests or that the public interest will emerge uncontested in policy and regulatory processes.[12]

The reality of the skewing of public–private boundaries towards digital platform interests in growth and profit is seen in the neoclassical analysis as a positive outcome. The shortcoming of neoclassical economic analysis is that it does not deal, other than in very limited ways, with the pivotal importance of power asymmetries in the way platforms generate and distribute economic value. Nor does neoclassical economics recognise the possibility of externalities that emerge from the aggregation of observations of individual behaviours made possible by platform operation. It can be helpful in assessing the likely outcome of alternative economic choices, but it is not intended to yield guidance on how to encourage digital platforms to operate differently in relation to public values.

The traditions in institutional economics and critical political economy analysis do consider both the individual and collective consequences of economic choices. They are concerned with developing evidence of platform market-related power asymmetries that lead to enfranchising only the few and to datafication processes that undermine public values. In these two analytical traditions, if policy and regulation are deemed essential or alternative ownership structures (including common property) are held to be desirable, the challenge is to establish how accountability to citizens, and workers as individuals or as communal group represent-atives, can be achieved. The contributions of self-regulatory measures, alternative business models and external platform policy and regulation are addressed in Chapters 5 and 6.

Chapter 4 turns to a discussion of the role of innovative technologies and practices, particularly the development and use of AI in the commercial datafication process, the implications for employment and markets for labour and the landscape of alternatives to the existing dominant platform business models.

Notes

1. Economic analysis is only one of many approaches in the social sciences that can offer insight. For others see DeNardis and Hackl (2015), Musiani et al. (2016), Parks and Starosielski (2015) and Plantin et al. (2018).
2. Readers familiar with economics will recognise that we use the word "satis-faction" for utility, a construct that requires assumptions.
3. The supplier is unable to affect the price by withholding supply because other suppliers will supply whatever amount is withheld.
4. The assumption that people take rational decisions (maximise their satisfac-tion) is challenged empirically by psychologists and behavioural economists (Kahneman and Tversky, 1979; Tversky and Kahneman, 1992). Employing these insights has led to ideas about how to influence behaviour (Thaler and Sunstein, 2009), how investors may come to have common beliefs (Shiller, 2019), and how people may develop cognitive biases or predilections (Bourgine, 2004; Kimball, 2015).
5. The implications of information for economics have been a major research area since Akerlof (1970), who observed that buyers of used cars did not have complete knowledge of their qualities. Because information, knowledge and beliefs are interrelated, the works cited in note 4 on behavioural and cognitive economics are also linked to the uneven distribution of knowledge among people.
6. For an early scholarly recognition of network economies, see Rohlfs (1974). The popular reference is to Metcalfe's Law, a claim that the value of a network is proportional to the square of the number of users. See also Rochet and Tirole (2003).
7. See Brautigan (1983) for this story.
8. See Arthur (2015) for the theoretical argument.
9. Monthly active users are those with a Facebook site who have logged onto their account in the last 30 days, see Facebook (n.d.).
10. In the US, newspaper advertising revenue declined from USD 20.4 billion to USD 13.4 billion from 2010 to 2017 and in the UK from £4.3 billion in

2000 to £1.7 billion in 2017. In the US, subscription and sales revenue was 35 per cent in 2017. Figures in this paragraph are from Statista (2019) and IAB Europe (2018). In the UK, digital advertising accounted for a 52 per cent share of a total £22.2 billion spend in 2017 as compared to the press 9 per cent, TV 22 per cent, direct mail 8 per cent, outdoor 5 per cent, with cinema and radio comprising the rest, see UK (2019e).

11. There is evidence that personalisation tactics have modest effects on whether digital platform users are exposed to diverse opinions (Flaxman et al., 2016). Those valuing diversity can be served by platform recommender systems that enhance the diversity of news sources (Bodó et al., 2019).

12. We do not develop a discussion of procedural means of making such choices. See Bozeman (2007) for a discussion of how public values failures can be considered. In the case of the media and communication industries, see Couldry (2019), Garnham (1997), Hesmondhalgh (2017) and Mansell (2002) for discussion of procedural approaches to choice-making, building on Amartya Sen's notions of capabilities and entitlements that are essential to a well-functioning democracy.

4. Technologies and datafication practices

4.1 Introduction

The nature of digital platforms as economic actors was examined in the first three chapters. We explained how these organisations transform existing business models in commerce and create new models for social media. We also highlighted some of the controversies and conflicts accompanying the expansion of platforms, enlarging our consideration of issues exposed by the development of platforms and their datafication practices. This chapter considers the business practices that make use of data generated by users and captured by platform owners to define a market for user attention. Central to this account are innovations in technologies variously labelled AI, predictive algorithms or big data analytics. The escalating application of AI by platforms raises further issues about how AI-enabled services will be delivered in the future. The most important of these issues is whether these services will eclipse the sale of advertising and e-commerce as the business models supporting platform growth and concentration. AI technologies rely on practices that extend beyond the passive capture of data to include methods of deepening and intensifying user engagement with platforms. This issue has immediate implications for competing business models for journalism content creation and distribution that we examine in the latter half of this chapter.

4.2   Why is AI related to platforms?

Platforms operate a multisided business model in which they offer free services to users, who are then exposed to advertising which generates revenue for the platform owners (as discussed in Chapter 3). Similarly, platforms selling goods and services to users can capture data about user preferences and habits with the aim of improving their ability to offer products and influence user buying behaviour. At first glance, efforts to attract an audience in order to sell advertising, or to better understand customer or citizen preferences and behaviour, appear to be extensions of the traditional activities of marketing and customer relations management. What distinguishes the way these processes are accomplished on digital platforms is a consequence of advances in the use of AI and machine learning. The processing of data using AI and machine learning techniques is a central feature of the broader "digital economy" in which the platforms operate.[1] The platforms as we define them in this book (see Chapter 2) are generating vast amounts of both user-contributed and user-interaction generated data. This data is being combined with many other public (government) and private (e.g. insurance) datasets to generate economic value. The companies involved include the platforms themselves as well as other companies in the data analytics, AI, blockchain, IoT, cloud computing and other internet services-based sectors. In this chapter, we focus on why it is essential to consider these innovations in the light of both their significance for economic value accumulation and for sustaining public values.

The achievement of AI was the aim of many of the early pioneers of digital computers. In the 1950s, the aspiration for research on AI was to discover whether characteristics of human intelligence such as learning, problem solving and heuristic formation could be convincingly simulated using computer hardware and software.[2] One definition of a convincing simulation was the Turing Test or "imitation game" in which humans communicating with an unseen interlocutor attempt to distinguish between the human and the AI. This defined the object of general AI and there are numerous illustrations of AI Futurism with respect to this broad understanding of AI. With a few exceptions – such as managing complex expert knowledge based upon extensive interrogation of human experts in order to reproduce their heuristics and patterns of association – expectations for general AI had singularly unimpressive outcomes despite more than half a century of research endeavour. Some researchers remain sympathetic with the aim of achieving comprehensive imitation or the reproduction of human cognitive and rational capacities or "intelligence", with applications that would include adaptive learning, sensory interaction, reasoned planning and creativity.

The advent of very large datasets – "big data" – has allowed for a different approach, often referred to as narrow AI. The aim is to use data to create systems that can reproduce or mimic the observed behaviour embodied in a dataset. Underlying this approach is a remarkable departure from previous understandings of how algorithms might be implemented in computer systems. Traditionally, algorithms were rules based, with finite, deterministic and effective collections of rules or steps for transforming inputs into outputs (Sedgewick and Wayne, 2011). In contrast, a data-driven approach to AI involves discovering the rules of the algorithm from data.

In the simplest example, a rules-based approach to solving the problem of what 2 + 2 equals is by establishing a rule on how to perform the addition operation between two numbers. A data-driven approach involves processing a large dataset containing many instances of 2 + 2 to create a system that will reliably report 4 as the answer. This may seem like "voting on the truth" and this would not be an incorrect summary of the implications. For example, if the dataset containing instances of 2 + 2 had a large number of 5s as an answer, the AI system might report that 2 + 2 equals 5. In this case, the AI system would *not* be like what either a human or a rule-based system would produce.

A more interesting example is the problem of constructing a sentence. Researchers can use a dataset with millions of sentences created by native language speakers to select how to put words together "like" the instances in the dataset. What "like" means differs depending upon the technique employed. Without human feedback or supervision, this approach to AI and the use of algorithms can find structure in large datasets and researchers can choose from the underlying algorithms, discovering those that best imitate human language behaviour. When the AI's discovery of structure or learning occurs without human intervention it is called unsupervised. Alternatively, humans may intervene and provide feedback as the AI system tries out various word combinations – hence, this is called supervised learning.

Hybrid methods involving supervised and unsupervised learning are possible means for decomposing the complexity of real-world problems. Decomposition of complex problems involving a hierarchy of operations or AI subsystems has produced the method of deep learning. Data-driven approaches to reproducing human behaviours such as sentence creation or language translation have proven powerful for achieving some of the aims sought by the earlier unsuccessful rules-based AI research agenda. Successful applications of data-driven algorithms include visual perception, speech recognition and language translation. Practical applications also include recommender systems (Amazon, Netflix), information search (Google), spam detection, and behavioural pattern detection used in financial fraud detection. Robotics tasks or AI-enabled drones for disaster relief operations also rely on these techniques. Sectoral applications include healthcare diagnosis, support for ride-sharing fleets, personalised financial planning packages, back office and customer-facing service automation and IoT applications. The wide scope of applications suggests that narrow AI may come to be viewed as a radical innovation in its own right in the future. Because of this potential, it is important to consider the first-mover advantage or head start that platform companies will have in delivering AI services in the future.

Innovations in data-driven AI techniques are employed by the platforms of principal concern in this book in multiple contexts from search to news aggregation, monitoring, forecasting, filtering and scoring.[3] However, data-driven approaches to constructing AI systems lack the systematic rule-checking available in rules-based algorithms. When an AI system has been created from data it is generally not clear whether it would produce different results if other, equally or more valid, datasets were used to construct it. The outputs or answers it produces may exceed or fall short of human capabilities. Nonetheless, data-driven AI systems are being used to answer very practical questions such as what the linkage is between user search behaviour and the type of product or experience being sought. Can the online behaviour of users make it possible to predict whether they are more likely to be interested in advertisement A or advertisement B?

Platform companies have been early adopters and large-scale investors in the scientific knowledge underlying AI and its practical application, and they are benefiting from the claimed predictive value of AI techniques. They have also benefited substantially from the availability of big data because this resource has been flowing freely to them as the result of the institutional rules that are now in operation. This applies both to user-generated data and to multiple "open data" repositories made available to the platforms under open data government initiatives at low or no cost. This has given the platform operators an important first-mover advantage in developing AI applications and, in turn, promises to reinforce other sources of market power stemming from the economies of scale resulting from the acquisition of this data (see Chapter 3). The platforms' use of AI applications extends beyond advertising markets to include efforts to restructure labour activities to achieve benefits from AI in their own organisations and to provide tools and resources for the application of AI throughout society.

4.2.1   Can AI predictions be trusted?

All these developments have implications for trust in AI applications. Whether AI predictions can be trusted hinges upon the discretionary or deciding power that AI-empowered systems are granted by humans. When AI systems are granted the main or sole responsibility to make decisions in human affairs, assurance is needed that these systems are reliable and accountable and do not bypass human rights to transparent processes and non-discrimination. In short, there is an expectation that AI system predictions should be trustworthy when, for example, they are involved in filtering information in a way that amounts to automating choices that humans would otherwise make. The expectation is that such information filtering should improve human agency and well-being, not diminish it. Yet, the AI systems that are used to provide online search capabilities are treated in neoclassical economic analysis simply as "two-sided matching" computational models (Varian, 2016). Often called prediction engines, the goal is to process data to yield reliable predictions, even though most of these systems are not based on statistical rules that allow estimation of error probability and range. Thus, a central issue in the trustworthiness of AI systems is the extent to which uncertainty governs the prediction process.[4]

Decisions based on the output of data-driven algorithms classify individuals, for example when an AI system is tasked with filtering employment applications. When decisions are taken based on these classifications, they affect people's life chances. If uncertainty is high, then there are consequences for the way the outputs inform decision making. The typical aim of the use of prediction engines is to produce reliable and fair results, "without bias".[5] Bias, however, arises from, and is inherent within, any social structure. Bias is reflected in data that is used in constructing data-driven AI systems. AI systems based on supervised learning "learn" patterns from training data that is incomplete and "unrepresentative". Human supervision also introduces biases, prejudices or ignorance. Decisions based on prediction engines can reinforce inequalities or infringe on consumer and citizen rights, for example, leading to race or gender discrimination in recruitment processes, to negative impacts for workers and their wages, or to the unfair treatment of vulnerable groups by social services. Multiple intermediaries may provide the data inputs for these systems, and those who must interpret the results have little or no capacity to correct for biases in order to avoid incorrect inferences (Barocas, 2014).

Innovations in AI systems and predictive models are designed to achieve selected goals and to meet performance metrics that are informed by judgements about desirable outcomes. Desirability is most often determined by those commissioning or constructing the systems (Selbst et al., 2019). AI applications development aims to translate human values into computational outcomes. For example, in the case of legal judgements, many judgements are routine applications of established precedent, but others involve non-routine judgements. Making this distinction is a challenging task for a data-driven AI system since the preponderance of data used to construct it will come from routine judgements (Pasquale, 2019). The uncertainty associated with these systems is raising concern about the use of data-driven AI systems to underpin decisions about access to public services, interpretations of surveillance data and the customisation or "personalisation" of online services. The biases arising from AI systems as methods of computation and classification are not new (Bowker and Star, 1999), but such methods are now being linked to discriminatory outcomes for particular classes and groups in an increasingly large number of contexts ranging from social support systems to policing (O'Neil, 2016). Despite their increasing "reliability" in some of these contexts, the regularity of prediction provided by AI systems is not the same as a claim to reliable outcomes from the point of view of equity or fairness (Helberger et al., 2015).

Data-driven AI supporting the platforms' systems encourages comparisons and choices. These may align well with the rules and norms of platform operators and those of other authoritative actors who benefit from the use of these systems. However, when these AI applications become entangled with, and shape, every aspect of life, then the lack of transparency of these systems is problematic. On the one hand, innovations in AI and machine learning are treated as signalling freedom from human bias and the promise of inclusive and equitable societies. On the other, innovation in these same technologies raises the risk of reducing complex social, political and economic environments to rules that foster multiple exclusions and inequalities.

4.2.2   Is there an AI "arms race"?

The capabilities afforded by AI are leading to the expansion of the number of actors who can carry out cyber-attacks and the range of potential targets. An "arms race" is developing between those seeking to defend against attacks, including the large platforms that, because of their prominence, are often favoured targets, and those seeking to perpetrate them. This race is a further spur to investment in machine learning and its applications.[6] The use of AI systems for fraudulent or malicious tasks that are otherwise impractical for humans is one of these new threats (Taddeo and Floridi, 2018). An example is "spear phishing", an AI-enabled cyber-attack that involves tailoring an e-mail to a specific individual, organisation or business, usually with the intent to steal data or install malware on a target computer or network.

Similar approaches are used in the promotion of mis-, dis- or mal-information within social networks when AI-empowered agents mimic humans interacting online. AI systems are also vulnerable to data-poisoning attacks (introducing training data that causes a learning system to make mistakes), adversarial initiatives (inputs designed to be misclassified by machine learning systems) or the exploitation of flaws in the design of autonomous systems. For example, cyber security is of increasing concern as IoT applications develop high levels of connectivity and machine-to-machine, as well as person-to-machine, communication (IoT Security Foundation, 2018; The Royal Society, 2018). With the threat of large-scale cyber-attacks growing, individual privacy and safety are increasingly at risk. In sectors such as healthcare, social media use, government reliance on big data and in data-driven AI systems used in the political sphere, risks to human rights and independence are substantial.

As AI systems scale up and exceed human capabilities in more and more fields, it is likely that cyber-attacks will become more prevalent and require preventive responses. The increasing ubiquity of AI systems is both a civil and a military matter. There are optimists about the outcome of these "arms races". For example, Facebook's chief technology officer says that AI systems provide the only means of preventing "bad actors" from taking advantage of platform services. He claims that they are likely to become so effective that in the next few years security breaches and illegal or harmful content will no longer be seen as having a "a big effect on the world".[7] However, applications in the defence sector, such as autonomous vehicles or drone aircraft and systems to counter misinformation, to deter or de-escalate conflicts and to aid command decisions heighten the potential for harm to individuals and groups. Applications that aim to protect the safety of citizens by identifying "risky" individuals or behaviours create possibilities for illegitimate bias, discrimination or compromises of human rights (Gangadharan and Jedrzej, 2019).

4.3   Will platform-based AI supplant humans?

The considerable attention devoted to AI in the press and by policy makers is driven by the perennial question of whether this new technology will replace jobs on a massive scale. Concerns about technological unemployment – the substitution of machines or other technologies for human workers – have waxed and waned throughout human history. Spikes in concern are signals of the introduction of new technologies with labour-saving potential. Closer examination of these periods of high concern is informative about the ways in which the relation between technology and employment is negotiated. Three type of issues are of particular importance: the level of employment, the location of jobs, and the job qualities.

Forecasts of future employment levels are nearly always unreliable because of the ever-changing nature of work and the flexibility of humans to master new skills and capabilities. Any job, or group of jobs, is not a static target but changes over time, as does the pool of those who might seek that job or type of job.[8] Further complicating employment forecasts are the uncertainties surrounding the speed of take-up of AI applications and the role of platform companies in delivering these applications as services. Employment forecasting typically focuses on "first order" substitution effects. Taking the predominant characteristic of a job, the question is whether future demand for this job characteristic is likely to rise or fall.[9] Because there are many jobs with a predominantly routine character, it is imagined that these jobs will be replaced by AI systems (including robotics) and the employment level will fall. This type of approach oversimplifies the nature of jobs and skills. Substitution of routine tasks by AI or automated systems makes it possible to reorganise multiple jobs in an organisation and creates complementary, as well as substitution, effects. Jobs are redefined as well as eliminated. Substitution of routine tasks also allows improvements in productivity which are likely to lower costs of production. Lower production costs make it possible to lower price or improve quality and, hence, to sell and produce more, creating a rebound effect in employment. These influences soften the impact of substitution for routine tasks. Nonetheless, in the OECD area, nearly 50 per cent of companies surveyed by the World Economic Forum in 2018 expected that automation would lead to some reduction in their full-time workforce by 2022, while 38 per cent of businesses expected to extend their workforce to new productivity-enhancing roles, with over 25 per cent expecting automation to lead to the creation of new roles in their enterprises.[10]

One reason for these expectations of automation effects is that AI systems make it possible to reorganise jobs that deskill and deconstruct them in favour of repetitive tasks that can be monitored by information systems. In earlier work, Zuboff (1988) raised this issue under the heading of the "information panopticon", a reference to older forms of worker surveillance that can now be performed by electronic systems. Those who foresee some disruption in the labour market as a result of the expansion of AI systems sometimes call for mitigating measures in the short or medium term to address structural unemployment (Pupillo et al., 2018).

Improvements in logistics including containerised shipping, and worldwide access to data communication using the internet, provide a framework for the global division of labour and, hence, the possible relocation of jobs. Lower-skill jobs migrate to lower-wage countries (or regions within a country with lower wages). Similarly, higher-skill jobs migrate to urban areas with larger numbers of highly educated workers, regardless of whether these are in London, Singapore, Shanghai or Mumbai. Despite recent reverses, the rules-based international trade system supports movement of goods and services with no, or relatively low, tariffs, and flows of data and information have been largely unimpeded by national boundaries. Some lower-skill jobs are routine in nature and provide opportunities for substitution of AI systems. However, these systems require major investment and are (so far) not as flexible as human workers, even workers with lower skill levels. In addition, the lower-wage countries have ample supplies of labour that offer an alternative to developing specific AI systems. Thus, at an international level, the impact of AI systems on employment levels is likely softened by the availability, lower wages and the flexibility of human workers to take on a variety of tasks.

4.4   Negotiating change

To take advantage of these new job opportunities in the digital economy as a whole, new skills must be developed, and particular areas of education must be emphasised. It is estimated that by 2022, around 54 per cent of all employees will require significant re- and up-skilling. It is also argued that as a relatively small number of "superstar companies" become increasingly powerful, workers will capture a smaller slice of the economic rewards, contributing to income inequality (Autor et al., 2019). Growing divisions between the rich and the poor have been attributed to the spread of advanced digital applications.[11]

As the capacity to understand, analyse and criticise data becomes integral to the functioning of the economy and democracy, the need for new skills is extending into every area of work activity. This is often articulated in policies seeking to promote STEM (science, technology, engineering and mathematics) subjects at all levels of education. However, improving and innovating services valued by users also requires skills combining those taught in STEM subjects with those taught in the arts and humanities, suggesting the acronym STEAM (science, technology, engineering, arts and mathematics).[12] Broader skills related to complex problem solving, creativity, critical thinking, people management and cognitive flexibility are also likely to be necessary. The challenge of combining these kinds of training is considerable.

There are also large gender and racial inequalities in the composition of the workforce that is being trained. Across the OECD countries, for example, men are four times more likely than women to be information and communication technology specialists and a pattern of under-representation is also present for racial and ethnic groups.[13] Local shortages of talented people at key places where demand is strongest, accompanied by barriers to migration (such as housing prices and resurgent nativism), mean that the demand for workers with these skills is outstripping supply in some countries, notwithstanding the fact that forecasts change rapidly depending on the source and on the timing.[14]

Ultimately, this change of direction in the nature of work reflects the power that managers have over workers in defining their job responsibilities and setting standards for performance. Given a choice between creating more meaningful and fulfilling work in which AI systems augment or improve work conditions, and defining work more narrowly that can be monitored and controlled to meet output targets, the latter path is being chosen by many organisations. The resulting jobs, being tightly defined in terms of function and output, can become "on demand" or "as needed", which gives rise to practices such as zero hours contracting or the "gig economy". These developments are not always seen as problematic by workers. For a variety of reasons, individuals may wish to have employment that is more flexible or does not involve as many hours or days of work in a week. Even if this is the case and there is a gradual change towards independent work, the more difficult question is whether these independent jobs, on balance, pay reasonable wages or offer benefits comparable to more regular employment.

This output or "piecework" style of organising "on demand" labour is consistent with the prevailing class, racial and gender inequalities that characterised pre-existing forms of labour exploitation, especially for low-income workers (Van Doorn, 2017). Flexible labour market optimisation, organised and managed by platforms, may be attractive to employers, but that flexibility often comes at a cost of reduced or absent employment benefits and uncertain incomes (Graham and Anwar, 2019). Studies of crowd-working, for example, suggest that it is often organised in commercial contexts such as Mechanical Turk or InnoCentive in ways that maintain substantial power asymmetries between workers and the owners or controllers of platform-accessible tasks, which increasingly are maintained by AI algorithms (Al-Ani and Stumpp, 2016).

As learning machines and algorithms extend their reach throughout society, new approaches to their governance may alleviate some of the potential problems. It is crucial to acknowledge, however, that "data and data-driven systems cannot claim all the credit for structural inequalities of an unjust society" (Gangadharan and Jedrzej, 2019: 896). In relation to the future of work and jobs, the challenge in the face of AI advances is not only to create new employment opportunities, but to create ones that will be valued by workers as "good jobs". Policies will be needed to support people who are having to transition to new jobs or face unemployment. If public values are to be respected, the analysis of the trustworthiness of AI applications and the future of work will need to move beyond the narrow framing of neoclassical economics to address issues of equity and inclusion, human autonomy and poverty reduction, and work will need to be treated as "the foundation of a modern moral economy" (Pissarides and Thomas, 2019: 7).

Digital content is key to the way most digital platforms generate revenues and profits. Content is central to the multisided business model since desirable content is what attracts platform users who become audiences that advertisers pay platform companies to access. We have seen already that AI makes it possible for platform companies to target and direct advertisements to specific users based upon observation of their behaviour and use of data that users provide. The logical next step from the perspective of platform owners is to tailor not only advertising but also content to specific users. We focus here on content that traditionally has been supplied by publishers and broadcasters with the aim of informing readers and viewers about events and people (news) and offering interpretations related to political, social and cultural choice (opinion); the traditional domain of journalism. We focus on this type of content because it is one where the platforms already have a major influence. Our consideration of journalistic content as a next step in platform development is only one of several next steps that follow logically from the capacities of AI to tailor and target information to users. It is possible to imagine next steps involving customisation of information about healthcare, education, social services and so forth. In each of these areas, analysis would involve different actors, processes of content acquisition and use, and consequences. Similar incentives and considerations would apply to these other areas. In the following we consider how journalism content is created, acquired and utilised, and then examine the issues and consequences of platform operation in this area.

4.5   Creating, acquiring and utilising content

By the beginning of the 2000s, there had been a shift away from free-to-air broadcasting media and the print media supply of journalism content towards platform provision. Compared with traditional means of reporting and editorialising, a key change was the proliferation of user-produced content by individuals posting news and opinion on social media websites. This phenomenon far exceeded previous means for individuals not allied with or employed by publishers or broadcasters to contribute journalism content. It provoked some commentators to question the quality of these contributions since the individuals contributing were not accepted members of the journalism community, though others saw this as a way of opening up to contributions by the public (Beckett and Mansell, 2008). Increasingly, however, self-published news and opinion became a publicly accepted part of the online environment. During this period traditional news publishers and broadcasters were engaged in creating online outlets for journalism. They attempted to translate the advertising-supported business model and, in some cases, subscriber-supported business models into the online environment.

These developments set the stage for platform owners to adopt an "aggregation model" as a means of gaining content. In a similar fashion to Google's creation of a search engine platform, platform owners could now create links to user-published and publishers and broadcaster-produced content, arranging these links by departments such as national and international news, sports, celebrity news, science, and so forth. Platform owners were exploiting a basic principle of digitalisation where "all of the media become translatable into each other – computer bits migrate merrily – and they escape from their traditional means of transmission … If that's not revolution enough, with digitization the content becomes totally plastic – any message, sound, or image may be edited from anything into anything else".[15]

As discussed in Chapter 2, platform companies saw opportunities relatively early in their development to shape consumer and citizen interaction online. The construction of platforms with content that would attract users was combined with customisation of content. This helped to create a unique user experience and to reinforce user participation or "stickiness" that would keep audiences engaged with varying degrees of success, increasingly exploiting AI systems to filter and make matches between content and observed user behaviour. Journalism content could be introduced in specialised news aggregation sites such as Google News, into the social network applications such as Facebook where interest groups proliferated, or on more specialised interest sites such as the UK's Mumsnet aimed at parents.[16] These communities of interest provided additional opportunities for targeting advertising and, of greater concern, opportunities to influence trends in fashion, political belief or social identity.

The traditional advertiser-supported business model favoured by commercial media used content to attract attention to advertising. Building on this, the platform strategy was to develop the means of predicting what content will intensify attention and to adopt practices of assuring that the "attention attractors" receive priority. Platforms use content to "catch eyeballs" or attention, and then to resell this attention to advertisers of commercial goods and services or political interests. Strategically, the platform operates in an "attention brokerage" market (Wu, 2016).[17] The platform approach to delivering journalism content was generally accepted by users. The reasons for this acceptance are complex. In part, the customisation of content to user interests meant that users could reduce the amount of time spent searching for preferred content. More controversially, the range of content available for aggregation and presentation to users had become more diversified and "less serious" in terms of journalistic values or ethics. Emotive stories such as those found in tabloid newspapers, human interest stories (often about celebrities) and partisan political editorialising became more prevalent as platforms bid for user attention. From a neoclassical economic viewpoint, these developments simply indicate that traditional publishing and broadcasting outlets were offering less desirable content, and that platforms were succeeding by offering content better attuned to user interests and better able to win their attention. However, both the methods and the results of these processes have consequences for the journalistic landscape.

4.6   Consequences

Content aggregation is the gathering of links across sources and their presentation in a format, order and variety responsive to explicit user choices or past user viewing behaviour.[18] Newspaper publishers and broadcasters argue that links to individual stories bypass (or displace) their decisions about assembling content in order to attract and retain readers or viewers of their products. Referring to platform practices as "aggregation" is strategic since it allows platform owners to claim they are not "republishing" or "copying" content which might infringe the copyright of newspaper or other media publishers. Instead, they claim they are merely providing users with a way to access information that more directly reflects their preferences (what they want to read about) and the newspapers can continue to offer advertising to readers on the linked pages.

Despite platform owners' claims, newspaper publishers and broadcasters have sought to curtail or be awarded compensation for aggregation practices. They argue that platforms unfairly benefit from their news production investment and that they achieve too small an economic return for their content to sustain the costs of high-quality (and especially investigative) journalism. While the publishers can also use these methods, they do not have access to other data about users under the control of platforms and, hence, are at a disadvantage. Legal disputes concerning aggregation practices generally have favoured platform practices although, as a matter of self-regulation, news aggregators have started to pay newspapers for content. Additionally, in Europe, legislation was amended in 2019 to address claims of copyright infringement associated with the platforms' use of "snippets" of news stories.[19] In the US, the platforms have benefited from the Digital Millennium Copyright Act 1998[20] "fair use" provisions that have been interpreted to mean that Google, for example, is not liable for copyright infringement when it displays copyright-protected "thumbnail" images without a publisher's consent. When there are modifications of the rules governing copyright in the digital domain, this is intended partially to rebalance the economic returns to content production with the returns to distribution.

Aggregation practices are affecting the operation of the digital content sector. These pressures include challenges to professional values, ethics and practices of journalism as journalists strive to maintain their audience.[21] Aggregation and other online journalistic content resources such as blogs or Twitter streams are weakening the link between the reader and the publisher (Newman and Fletcher, 2018). At the same time, the digital platforms are where news is increasingly sourced by readers (particularly younger readers).

There is a strong basis for linking media content, especially when this involves news and editorial content, to public values. The goals of safeguarding media diversity and pluralism inform regulations that are aimed at fostering the production of media content containing diverse viewpoints and enabling these to be found when platform users go online (Helberger, 2018; Hesmondhalgh, 2019b). As the Council of Europe puts it, media freedom and pluralism "are central to the functioning of a democratic society as they help to ensure the availability and accessibility of diverse information and views, on the basis of which individuals can form and express their opinions and exchange information and ideas".[22]

These developments are viewed differently by the three economic traditions. A neoclassical economic analysis sees these developments as affording virtually unlimited variety and little or no constraint on choice. However, a neoclassical analysis may still recognise a "public good" argument for "merit goods", goods for which there is a socially justifiable need, but an absence of a business plan for supporting their production.[23] The neoclassical framing, however, dramatically reduces the *a priori* justification for journalistic expression as a merit good in a world apparently lacking in scarcity of journalistic expression. It suggests that those arguing for provision based on market interventions of various kinds should demonstrate deficiencies in the plurality and diversity of offerings that are available in the marketplace. In its starkest form, the question is why society should empower a specific group or groups to represent the public interest when that interest is best revealed by individual choice about what media content is consumed. A corollary is that valued content is expected to attract enough audience to receive sufficient advertiser support.

In an institutional economics framework this line of analysis is regarded as reductionist in that it denies the legitimacy of arrangements that are collectively or democratically agreed to serve a public purpose. Market-driven variety of content production, even variety endorsed by choice, is no guarantee that public values will be upheld. For institutionalists, the structure created by the attention economy is depicted as antithetical to the reasoning that gave rise to representative democracy (as opposed to common plebiscite) or to free expression (a means of protecting the individual expression of the minority). In an institutional economics approach the construction of organisations seeking to discover the public interest and to articulate public values is acknowledged as an arduous process. It is subject to capture by special interests and it is at risk of demagogic outcomes. This approach does, however, emphasise the legitimacy of mandates for media services that are responsive to the public interest and purpose.

Supporting this view is the vast amount of online expression which is intent on causing distress and, in some instances, leads to physical harm, targeting those with few resources to which they can turn for protections in countries around the world. Neoclassical economics usually assumes "free disposal", which means that the recipient of these messages can simply ignore them, an assumption that is difficult to make in many contexts, e.g. young people's use of social media. Even though platforms create opportunities for contributing to and accessing content that differs from the agendas set by elite media producers, they can exercise gatekeeping power because they control content discoverability, searchability and visibility (Schlosberg, 2018).

Critical political economy analysis is helpful in evidencing this history. From this perspective, the media industry enables the interests of capitalists to prevail, exacerbating social and economic asymmetries of power. Innovations in the earlier period of digitalisation of content were greeted as bringing "more menace than promise" (Schiller and Miège, 1990: 166). Advertiser-supported content production and the machinery of the attention economy were examined as parts of the process of consolidating power by elites who seek to influence audiences to engage in commodity capitalism. In the face of digital enthusiasts, opportunities for self-publication and online social communities, in this tradition, the emphasis was on how the changes in the media industry were "marked by the inequality of exchanges" (Mattelart, 2000: 107). The effort by the platforms to cluster audience attention to sell to advertisers is seen as being aimed, ultimately, at eradicating human behaviour that is not calculable and subject to prediction using algorithms in the interests of the capitalist owners of the means of content production (McGuigan, 2019).

A critical political economy analysis of contemporary content production and use emphasises business models as embodying dominant class interests and, more fundamentally, raises the important question of whose interests are being treated as consistent with the "public interest". In this view, digital content inevitably is shaped and dominated by those owning and controlling the commercial platforms. Although platforms have risen to dominance through individual actions, those in control of the means of content production are strongly influenced by private interests. Rather than concede the inevitability of a system dictated by private interest, the need for public ownership and control or, at the very least, strong public governance, is asserted.

Institutional economics and critical political economy frameworks draw attention to the complexity of media content as a "social and cultural practice" (Freedman, 2008: 157) in contrast to neoclassical individual "choice". Arguments about the complex relation between the media content industry, and especially news production, and public values and purposes will persist. They are influenced by past struggles and by the institutional norms and rules that these struggles produced as accommodations or compromises. For analysts working in the institutional and critical political economy traditions, the neoclassical economics focus on the potential of digital technologies and the capacity for self-publication to facilitate democratic freedoms downplays the importance of structural asymmetries of power and their consequences on the supply side of media markets as well as the diversity of "users/audiences" (Benkler et al., 2018).

4.7   Conclusion

The breadth and nature of the consequences of AI system applications are uncertain, but the tendency remains strong to treat all forms of digital innovation as "the biggest technological juggernaut that ever rolled" (Freeman and Soete, 1994: 39). This view calls for consumers and citizens to "adjust" to the shock of the platforms' use of AI-systems as an enabler of radical platform innovation and as an opening for an ever-larger role of platforms in delivering AI applications. What the consequences of AI investment and the digital platforms' use and supply of these systems are for economies, and for the capacity to uphold public values, depends upon whether applications are for better or worse. AI developments raise questions of ethics (addressed in Chapter 5) and the uses of AI cannot be divorced from wider questions about their contributions to economic growth and productivity or to disadvantage and inequality.

As commercial datafication practices continue to shape the digital content industry in Western societies, the question is what the implications are for public values when platforms operate with incentives to manage the attention economy. Those seeking expression can enter the online arena to communicate messages, mobilise public opinion and engage in civic discourse. Those with little sympathy for public values have free rein to seek user attention. Appeals to self-interest and the cultivation of fear, hate and loathing have equal standing with appeals to public interest or rational discourse in seeking user attention.

The outcome of this contest for attention depends on one's view of human behaviour. The danger is that greater attention may be given to those who shout the loudest, make emotive and unsupported claims, and employ the tools of demagoguery. If social exchange or "sharing" in pursuit of cultural, educational, religious and political goals had been given as much encouragement as commercial datafication, the pattern of jour-nalistic content development might have been different. Much greater emphasis might have been given to non-commercial content in support of democracy. Means of resolving these conflicts are available through self-regulation and alternative business models (discussed in Chapter 5) and through external policy and regulation initiatives (discussed in Chapter 6).

Notes

1. It is estimated that the size of the "digital economy" is in the range of 4.5 to 15.5 per cent of world GDP. Using Global Internet Protocol traffic as a proxy for data that is collected and analysed by platforms and other components of the digital economy, it is estimated that this will grow to 150,700 gigabytes per second in 2022, up from 45,000 in 2017. Using UNCTAD's definition of platforms, the market capitalisation of platforms greater than USD 100 million was estimated at USD 7 trillion in 2017 (UNCTAD, 2019).
2. See Solomonoff (1956: 1), referring to McCarthy, Minsky, Rochester and Shannon.
3. Media play an important role in increasing expectations about AI, with news articles dominated by industry products, initiatives or announcements, see Brennen et al. (2019).
4. For challenges in developing machine learning and AI-based computational systems, see Ghahramani (2015).
5. See UK (2018b: 18).
6. For a succinct overview of the technology arms race arguments, see MacKenzie (1989). The term "arms race" has been used in neoclassical

economics treatments of innovation and competition (Baumol, 2004) and in discussions of digital services innovation (Noam, 2019).
7. See Simonite (2018) citing Mike Schroepfer.
8. There are, of course, exceptions, e.g. it is not a rapid process to become a neurosurgeon.
9. See, e.g., Berger and Frey (2016), Brynjolfsson and McAfee (2014), Davies and Eynon (2018), Frey (2019), Frey and Osborne (2017), LaGrandeur and Hughes (2017) and Marcolin et al. (2016) for examinations of impacts on various kinds of jobs and tasks.
10. See WEF (2018).
11. See Keen (2015).
12. See NESTA report, Siepel et al. (2016).
13. See OECD (2018) and National Science & Technology Council (US, 2018b).
14. See, e.g., EC (2019b), Manyika et al. (2017) and Schwab (2019).
15. US (1990: 4) citing Stuart Brand, and see Ibrus (2019) on cross-media developments.
16. See https://www.mumsnet.com/ (accessed 2 January 2020).
17. There are different ways of interpreting this market, see Feld (2019).
18. See Steinmueller (2003).
19. See European Parliament (2019).
20. See US (1998a).
21. See Eide et al. (2016), Fenton (2019) and Zelizer (2017).
22. Council of Europe (2018: 1).
23. See Samuelson (1954: 387) for the public goods argument, Musgrave (1959) for the merit good concept and, in the specific case of media, Freedman (2008).

5. Self-regulation and alternative business models

5.1 Introduction

Platforms are an increasingly dominant form of organising social, political and commercial relationships. In large measure, platforms are a realisation of earlier visions promising unprecedented freedom to access information, a cascade of entrepreneurial opportunity and connections with people throughout the world. Realising any vision has unintended social, political and economic consequences as acknowledged by each of the strands of economic reasoning. Neoclassical economics maintains a faith that negative consequences are remediable through competition and awaits definitive confirmation that competition is fundamentally impaired. Institutional economics and critical political economy share an expectation that some of these consequences are persistent. They are treated as the structural result of pre-existing rules, norms and standards that opened a space for platform developments. In this context, societal expectations for these platforms involve the freedom to express ideas and to communicate with those we choose, well-defined rules to prevent abusive and violent behaviour, and content with informational, educational and entertainment value (not always in that order) for all members of society. Underlying these expectations are basic ethical principles such as fairness, transparency and honesty.

When the platforms acknowledge negative unintended consequences associated with their operations, they often claim to be taking mitigating actions. In this chapter we identify some of the virtues and shortcomings of self-regulatory strategies and examine efforts to introduce ethical practices. We find that the shortcomings of these efforts indicate a need to consider alternative platform business models.

5.2　Platform self-regulation strategies

Self-regulation involves norm- and rule-setting through internal decisions. Platform self-regulation has enabled a small number of companies to grow substantially by successfully exploiting the business models discussed in Chapters 2 and 3. Their success has put them in a strong position to argue against new externally imposed regulation. Prominent strategies are a drive towards global market expansion, strategic use of standards to establish a strong market position, private establishment of service terms and conditions, the exercise of control over content moderation processes, and lobbying against external oversight of their businesses. In addition, the platforms are incorporating ethical codes for the use of AI-based technologies and applications as a form of self-regulatory practice.

5.2.1　Global market expansion

The dominant platforms strengthen and diversify their revenue streams by expanding globally, regionally and locally. They develop platform-owned and -operated applications and services suitable to specific markets and acquire and assimilate applications and services produced by other companies. When it is helpful for acquiring users, they "localise" their practices and content to local languages, cultures and political conditions. Their central claim is that global expansion involves delivering the benefits that platforms offer to an ever-increasing number of users. Examples include inter-user communication (social media), e-commerce, and ways to access information on the internet (search). The expansion strategies involve opening or deepening markets that are not yet saturated, which may involve investment in the local connectivity infrastructure. Expansion also involves diversification into lines of business complementary to the initial source of their success, e.g. games, messaging services and streaming media (Netflix has vigorously expanded globally based on a subscription model, partially employing platform principles).

Two criticisms of the strategies of global expansion are displacement and foreclosure effects. The displacement argument is that the large scale of some platforms and their ever-expanding array of service offerings crowds out potential or emergent competitors, particularly those originating in local or regional contexts. It is difficult to assess the significance of displacement effects because platforms that might have emerged in the absence of these global expansion strategies are not present. Some insight nonetheless can be gained by considering displacements of pre-existing actors in the digital landscape. The displacement effects that challenge traditional newspaper and broadcast news industries and the difficulty publishers and broadcasters face in consequence were discussed in Chapter 4.

Foreclosure effects refer to the suppression of alternatives. The most common mechanisms for achieving foreclosure are mergers and acquisitions. Foreclosure is an interpretation of what happens when a new company develops an alternative to a platform's practice or technology. Foreclosure begins when this event is followed by a business merger or acquisition of the company responsible for the alternative with an existing platform company. Typically, after merger or acquisition, this alternative is either integrated into the platform offerings or shut down – in either case, foreclosing independent development of this alternative.

Examples of mergers and acquisitions that have been integrated with existing platform functionality, or shut down, include Google's acquisition of DoubleClick, an online adMaker, as well as Waze (a GPS-based app) and YouTube. Microsoft acquired Skype and LinkedIn, partially integrating them with its other offerings. Amazon bought Quidsi, an online retailer that owned Soap.com and then shut it down. Facebook bought Instagram, an anonymous messaging app service, and continues to operate it alongside its principal offering (although there is speculation that it will integrate it in the future). It is often claimed that the reason for shutting down a service is a lack of utilisation. However, mergers and acquisitions may end whatever prospects potential competitors might have had for growth, limiting the extent to which the acquiring platform faces effective competition. The rejoinder is that such growth prospects were less attractive to the owners of the alternative service than the terms offered for a transfer of ownership to the acquiring platform.

For platform owners, global expansion turns on the claim that this expansion is evidence of the value that companies bring to their users. Without these users, platforms would be unable to grow as rapidly or incorporate innovative features. It follows, from the platform's perspective, that what they are doing successfully in attracting customers is in the interest of millions of people. The claim that there is an unserviced interest associated with public values then appears to be special pleading by those who are being bypassed or ignored. This claim is reinforced by the observation that platforms offering opportunities for user expression are open. Therefore, it may be claimed that they cannot be exclusionary. When some forms of expression elicit lower levels of engagement, this is seen as a problem in making effective expression, rather than being due to platform practices.

5.2.2   Strategic uses of standards

In the digital ecology, technical compatibility standards are required to manage the exchange of data between platforms and their users (David and Steinmueller, 1994). Such standards are also necessary for managing the data retained by platform owners and for monetising it, either by selling it or offering services based on the data to others. Technical compatibility standards not only facilitate data collection, retention and transfer, they are a means to advance the strategic interests of platform owners. Many of these standards are set by platform owners and collaborators under self-regulatory arrangements that reflect the platform owners' strategic interests and those of their suppliers and advertisers.

Regarding users, for example, it is conceivable that standards could be set in ways that make data entered by users more portable across platforms. Doing so, however, would make it easier for users to exit one platform for another and so such standards are unlikely to be voluntarily created by platform owners. Some platforms such as Facebook and Google do make it possible for users to view the data they have generated in the process of interaction with the platform. Going further than this to enable ready portability of user-generated data neglects the fact that these data would not exist in the first place without the services offered by the platform. It would also mean that an important source of revenue for the platform owner is bypassed. Self-regulation by platforms involves control of user data and compliance with any legislation that limits the reuse or transfer of this data. Standards that would enable users more easily to transfer data, often labelled "open data" initiatives, could create the possibility for third parties to exploit this openness to their own advantage. If this were to happen, the platform owner might be viewed as complicit in enabling this form of exploitation.

Technical compatibility standards are essential for constructing the means to monitor user behaviour which is central especially to the advertiser-supported business model. A key feature used by platforms and other web service providers is the "cookie", a small amount of data that is stored on the user's computer when accessing a web page that is maintained by a platform or other site. Cookies allow the web server to record repeat visits to a website and to retain information about user authentication. Cookies facilitate the placement of ads relevant to a user and other means of managing the user experience in the use of websites, including those managed by platforms. The governance of cookies is a complex mix of self-regulation, regulation by civil society organisation (e.g. the Internet Engineering Task Force) and legislation.

A fundamental issue with "cookie" standards that facilitates user platform activity is that they can be used for purposes of user surveillance. Sometimes this is for purposes of direct value to users such as maintaining their credentials to access information. At other times, it is for purposes of indirect value to users such as summoning advertising that might be of interest. And, sometimes it is for compromising the privacy of users' online interaction. In the latter case, although most users are aware of messages informing them of the use of cookies and other means of monitoring their online behaviour, a survey of those aged 18 and over in the UK found that 47 per cent of respondents (sometimes 35 per cent, or always 12 per cent) clicked messages offering to inform them about how advertising is personalised for them, whilst 53 per cent never did. Of the 35 per cent who sometimes looked at these messages, 50 per cent reported that they simply ignore advertisements targeted to them using cookies. Among the same respondents who sometimes looked at such messages, 29 per cent indicated they did not believe they could do anything about them.[1]

One issue raised by these kinds of behaviours is digital literacy. Under self-regulation, offering a user a capability for control is taken as equivalent to addressing the issue of user control. If relatively few users are prepared to exercise control, however, this equivalence is questionable.

This poses a dilemma for the platform owner. *Requiring* users to exercise control will have a negative impact on those who are indifferent to control (for whatever reason). In practice, a minority of users may have an active interest in extending their control of platform surveillance tools like cookies. The argument is that offering control therefore addresses the issue of control *for those who care about it* and, hence, the issue of user control is addressed. However, this does not address those who wish for control but are unable to understand and act on the capability for control that is offered. This becomes an even larger issue if many of the standards adopted are, in effect, means for controlling the user. Self-regulation rarely involves commissioning expertise or empanelling representatives of the user community with a critical view of platform operations. What the platform owner decides is best for the user in terms of a standard becomes the standard that is deployed.

5.2.3   Service conditions and codes of practice

Platforms not only are able to exercise their own judgements by defining the architecture of software and hardware through standards, they also serve as private "regulators" of user interactions with platforms. This is because platform owners control the terms and conditions offered to users. These terms and conditions may be changed to reflect new opportunities for datafication and data circulation. For example, when Facebook announced a new "privacy-focused vision" for its messaging functions in 2019 in conjunction with its integration of Instagram, Messenger and WhatsApp, this was presented as giving users "clear control over who can communicate with them and confidence that no one else can access what they share" using end-to-end encryption.[2] However, by integrating these different functionalities, Facebook can use the meta-data (e.g. time, origin and destination) of messages to create an even larger repository of information about social connections within its user community than it previously held.

Digital platform companies are developing and, in some instances, making their codes of practice public. These extend to the ethical use of AI and machine learning for collecting and processing data (see section 5.3) and these codes are translated into their terms of service agreements. For example, agreements might specify what forms of unlawful and harmful content are restricted. The argument favouring self-regulation is that users are fully (or at least reasonably) informed and that they can make rational choices on that basis which enables them to regulate their own contributions and readership behaviour.

The platforms may also use contractual "most favoured customer clauses" to control their market relationships with other companies in the business ecology (e.g. Amazon, iBookstore, Booking.com or Expedia). Such clauses typically commit a seller not to charge a lower price on one platform than it does on another. It may also restrict the discounting of sales directly to customers (i.e. preventing sellers from charging lower prices at their own sites). These practices can promote a platform as the best outlet from which to purchase a good or service and restrict the potential entry of other platforms offering discounted prices (e.g. by sharing some of the seller listing cost with buyers). Such self-regulatory strategies are designed to maximise a platform's control of its core asset (data), while, at the same time, creating incentives for providing innovative products and services.

Platform owners' intentions and abilities to set the terms and conditions of platform use and codes of practice in their operation that meet social expectations for fairness and transparency vary, but, in most cases, the consequences of these terms and conditions are not at all transparent to users. This is because of the asymmetry between the information available to users and the information that platform owners have about the consequences of agreement. As with compatibility standards, the unilateral decisions taken by platform owners are what determine the result. There is, additionally, a dilemma concerning user indifference, similar to that regarding compatibility standards. Within the journalism and user communities there are, of course, knowledgeable individuals who do analyse and offer editorial opinions about terms and conditions and codes of practice, but these are consulted by a minority of users. It would be far-fetched to conclude that negative commentary leads to substantial change in user choice about which platforms to join. This is, of course, subject to change. As we discuss in the next chapter, the same forces that give rise to "regulation by outrage" can produce negative reputational effects for platforms, perhaps with larger consequences for these practices.

5.2.4   Strategic content moderation

The problem of defining a boundary between illegal and permissible content or the extent of harm created by legal content requires judgement. To increase their awareness of the content and communications on their platforms, platforms need to trace users who contribute content with the risk of charges of privacy invasion. In response to concern about the social harms associated with illegal and harmful content, the platform strategy is to attract positive attention to their policies and to emphasise their investment in content moderation practices involving blocking or promoting content. On their own (or in response to new legislation), the platform owners are tightening their definitions of hate speech, hiring more content reviewers and, sometimes, establishing content advisory groups. They are also employing content credibility signals such as YouTube's showing of Wikipedia and Encyclopaedia Britannica information next to its videos. They also remove content when they are notified that it does not meet their own or legal standards, such as prohibitions of user posts that infringe copyright or contain illegal or harmful content.

These self-regulatory practices mean that the platforms are making private decisions on behalf of the public. Nonetheless, they are accused of being insufficiently proactive, for example, responding only when advertisers threaten to withdraw their advertising when media coverage threatens their brand, or there is a public outcry provoking legislation to oversee their practices. Responses to criticism of their content moderation strategies include YouTube announcing that it would ban comments on videos featuring minors to address parent and advertiser concern about the use of its platform by paedophiles. Facebook announced it would limit the reach of content spreading anti-vaccine misinformation after exposure in the press and an intervention by the Chair of the US House Intelligence Committee. Facebook, YouTube, Spotify and iTunes dropped content from INFO*WARS*, a channel devoted to conspiracy theories. Twitter and Google both have introduced changes in the way political commentary occurs on their platforms. Nevertheless, the dominant digital platforms frequently are accused of failing to moderate according to the standards they set in their own guidelines and terms and conditions.

The platforms claim that their content moderation strategies are not editorial functions, but self-regulated moderation faces the challenge of filtering out false information, hateful or otherwise harmful content and deliberate misinformation, without impeding free expression. Moderation is mostly undertaken after publication because it is assumed that prior vetting would be burdensome for users. Hence, a consequence of open platforms and user-produced content is that damaging content will be posted before it is filtered. The platforms rely on algorithms and content flagging with some human moderation and a variety of other tools and practices to improve credibility signalling and to filter out malicious or false content. Their aim is to strike a balance *of their choosing* which keeps users on their platforms while maximising opportunities to monetise their data.

With a view to assuring governments and citizens that their content moderation practices respect public values, a platform may cooperate with content producers. For example, Facebook is working with the press and other organisations to undertake fact-checking and to filter news flagged by users prior to elections. This strategy can place a burden on professional content producers, and it may require newspapers to prove that content is false for it to be removed or accompanied by a comment. Another strategy used by Google, Facebook and Twitter is to support not-for-profit websites such as First Draft News which allows users to submit questions to help verify the accuracy of content. Yet another is to ban certain types of content altogether, as when Twitter or Spotify announced they were banning political advertising. In such cases, the platforms may be accused of being *too* proactive, leading to the suppression of speech or censorship. They are also charged with turning a blind eye to very poor working conditions for an outsourced workforce that is tasked with implementing their content moderation policies in locations around the world.

5.2.5   Responding to external regulation

Platforms make numerous strategic moves in response to the threatened or actual introduction of new regulations. Their communication strategies aim to promote their commitments to public values. For example, "Google cares deeply about journalism – We believe in spreading knowledge to make life better for everyone. It's at the heart of Google's mission."[3] In cases where the platforms have been held to account by legal actions bringing charges of infringing data protection legislation, this is resulting in fines, some of which are substantial. These have yet to compromise the growth trajectories of the largest platforms.[4] Another strategy is to settle cases out of court with relevant authorities and to agree to transparency audits of their operations, for example, regarding their use of "personal data" as defined by law.

The industry view so far is that if remedies are needed to increase transparency and accountability, these principally should be the responsibility of the platforms. Alternatively, it is argued that better solutions to potential and actual harms will be provided by innovative technologies in the future. In the face of increasing scrutiny (which we consider in Chapter 6), it is unclear how far the platforms will succeed in denying or deflecting criticism by using lobbying, communication campaigns, tactics to draw out court proceedings, and incomplete (or incorrect) responses to governing authorities when they demand transparency. The platforms generally take an adversarial position on external regulation, translating their market power into political pressure to forestall or delay implementation, or they try to shape regulation to their advantage.

5.3   Ethical principles and practices

All the instances of self-regulation strategies discussed in the preceding section are shaped or inflected by implicit or explicit intent to conform to ethical principles and practices. Intent does not necessarily guarantee ethical outcome, in part, because of contested claims about what is ethical. For example, an ethical standard which maintains that platform users should experience strong personal privacy protections can interfere with the business objective of serving their needs to be informed, educated or entertained, which can also be considered an ethical standard.

Ethical principles and practices offer a bridge between self-regulation and external regulation. In principle, if platform companies were to bind themselves to acceptable ethical principles, and observe these in practice, there would not be a need for external regulation. There would remain the question of how social acceptability is established and what should be done if a platform does not adhere to these principles. There is no shortage of pronouncements on ethical principles, although, so far, there is little machinery for ensuring these principles are adhered to. This is why an ethics discussion is located in this chapter rather than in the next chapter, which considers external policy and regulatory proposals.

Ethical principles that may be reflected in the platforms' codes of practice are informed by professional standards, technical capabilities and social norms (Ananny, 2016). They address who or what is to be held accountable and to whom. Self-regulation presumes that platform owners will embed ethical principles as a matter of course. In doing so, however, they have little interest in revealing their hard-won knowledge about the effectiveness of their strategies and practices in generating revenue. A direct conflict between ethical standards and transparency and commercial interest is often present. Moreover, considerable confusion exists about the feasibility of transparency in the case of platforms because a complex system of norms, rules and techniques interacts in ways that are often not transparent to their creators and practitioners (Ustek-Spilda et al., 2019).

Advancing applications of AI and machine learning in use by the platforms are raising the profile of efforts to devise ethical principles and translate them in the design of digital applications. The idea that technology developers who advance the capabilities of technological systems have an obligation to consider the societal consequences of their innovative practice is a long-standing one, even among economists, despite denials in some instances. Herbert Simon, noted American economist, political scientist, cognitive psychologist and AI pioneer, observed that "with respect to social consequences, I believe that every researcher has some responsibility to assess, and try to inform others of, the possible social consequences of the research products he [sic] is trying to create" (Simon, 1991: 274). In Chapter 4, we discussed developments in AI. Many initiatives that are underway address cybersecurity and safety. These are producing biases when AI-related algorithms are used to supplement, and even replace, human decision making. Nevertheless, there is a long history of excessive optimism about the capabilities of AI. For example, Simon predicted in 1957 that computers would outplay humans at chess within a decade, an outcome that actually required 40 years to occur.

Multiple sets of ethical principles and codes of practice are being devised, with some 80 statements voluntarily recorded at a website in June 2019.[5] Approaches differ depending on commitments to ethical perspectives, from consequentialism to deontology to virtue ethics, and on how these can be put into practice (Sullins, 2016). Floridi (2019) notes that this proliferation creates multiple risks including "ethics shopping", corporate efforts aimed at favouring self-regulation and corporate relocation to countries with less onerous ethical standards. Multiple actors are involved in ethical guideline development on the United Nations level, regionally and nationally as well as by industry, professional associations and civil society organisations.

The application of ethical principles and commitments is uncertain as far as outcomes are concerned. The ethical development of the IoT, for example, involves addressing "fixable" ethics-related problems to enable implementation of a new application, often with little attention to social and economic inequalities that an application may give rise to. Security-by-design and ethics-by-design initiatives are discussed and, in some cases, implemented, but internet connected devices are being sold lacking even the most basic safety provisions. Additionally, conflicts of interest are present when the same organisation is responsible for the development of AI applications and their ethical use. These conflicts raise questions about whether it is necessary to regulate AI and the use of AI by platforms. Such regulation could aim to balance market growth with accountable ethical practice with respect, for instance, to data protection, privacy and the displacement of labour. The profusion of efforts in this area suggests that, at a minimum, self-regulation needs substantial broadening and deepening to meet social expectations.

5.4   The alternative platform landscape

There is a landscape of business models that provides alternatives to the advertiser-supported, mass participation and AI-empowered platforms which are the principal focus of this book. These models are shaped by the pre-existing rules, norms and standards that provided an opportunity for commercial platform development, by the strategic actions of platform companies and by user choices. We have emphasised that platform growth is supported by rules and norms allowing platform owners to appropriate data provided by users. This growth is also supported by technical standards that allow platforms to observe user behaviour, gathering even more data about user interests, interactions with online resources and connections with other users. The private ownership of this data provides platform owners with a substantial advantage in the competition for user participants and their attention. This ownership norm represents a definition of a boundary between public and private, favouring corporate initiative and control. How does this process of garnering competitive advantage look through each of the three economic lenses?

Neoclassical economic analysis assumes that private supply is most efficient. Under present arrangements, both personal data and data gathered observing user behaviour are treated as if they are a "natural" resource available for appropriation for commercial gain.[6] Following this logic, the private supply of platform services is regarded as the most efficient means of organising their services (UK, 2018d). This assumes that data can flow freely to private companies, unencumbered by constraints other than those for intellectual property protection. The platforms' right to the use of data, and their bundling of zero-priced content and curatorial services with positively priced advertising and data collection, give rise to "winner-take-all" competitive advantage. In this analytical frame there is very little scope for alternatives. The incentives for the maintenance of the predominant business model are strong, not the least because of the economic value of data. In the US, personal data was estimated at USD 76 billion annually in 2018, rising to USD 198 billion by 2022,[7] and these figures do not include the value of data gathered by observing users' online interaction.

In contrast to neoclassical economics, institutional economic analysis assumes that reforms in platform practices or in the structure of markets may be needed, depending on empirical assessment. A critical political economy analysis assumes that public or collective supply of platform-related services is consistent with the emancipation of exploited individuals, classes or groups. For this reason a shift towards public service provision is much more likely to secure public values. The problems created by the predominance of platform commercial datafication raise questions about whether a change in the boundary between public and private platform provision is necessary to secure those values – questions much more likely to be addressed in an institutional or critical political economy framework.

5.4.1   Changing the public–private boundary for data control

In order to enlarge the space available for non-commercial alternative platform provision, different rules and norms for the ownership and control of data need to be considered. In this section, we describe initiatives that have been proposed or undertaken to make a change in the public–private boundary. Some of these initiatives are limited because they lack affirmation or they have not achieved the legitimacy of a policy measure which, in some cases, would involve market intervention and new legislation. These kinds of interventions are examined in Chapter 6.

Within the existing framework of rules and norms governing the control of data, initiatives are limited to using technical features of the internet to interfere with the datafication processes employed by platforms. For example, it is possible for users to "opt out" of datafication by using "anonymising" internet connections and denying the placement of cookies on their machines. Some of those who use these techniques are motivated by a desire to maintain their personal privacy and prevent surveillance. The fact that this will prevent them from benefiting from the customisation of content is outweighed by their desire for privacy.

Others, however, are engaged in attempting to imagine and experiment with a different ordering of data ownership and use. Their aim is to re-appropriate ownership of their identity and data with a view towards negotiating a different settlement of ownership and control that favours the private individual. Projects such as Solid and the Hub of all Things,[8] for instance, aim to enable people to extract value from "their own data" should they wish to do so by trading their data; and this is sometimes associated with "data dignity".[9] This approach suggests a change in ownership rights, not only in data that users provide directly, but also in data generated by their online interactions or derived from the platforms' observations of their online behaviours. However, the tools being proposed are limited to enabling individuals to choose where to keep their personal data and by whom it should be used. This type of approach might lead to the creation of one or more "identity management" intermediaries providing a focal point for users to learn and make decisions about the use of the data they provide or that is linked to their name.

A related set of initiatives, implemented as pilots at the time of writing, is the creation of "data trusts".[10] These involve entrusting personal data to an organisation with the authority to negotiate access to this data on the user's behalf. Data trusts are an extension of identity management intermediaries. The implication is that users will make numerous decisions about what data to share with which type of third party and this requires a relatively high level of digital literacy. Data trusts can still facilitate commercial datafication, but in a way that is claimed to be subject to the users' control. Data trusts can be interpreted as a bid to create a new market for data. It remains to be seen whether commercial actors, such as platforms, will be willing to become customers of data trusts. Implicitly, data trust organisations are founded in anticipation that platform methods of gathering the data they hold on behalf of users will become onerous or expensive, opening a market for their services.

Going beyond these initiatives to consider more fundamental changes in the boundary between public and private (individual or corporate) data control would require legislation to change the rules underpinning data ownership and beyond the changes introduced by existing privacy legislation. For example, an intervention might be imagined that involved paying users for the use of their data, although how and by whom the value of this data would be determined would be a major issue. If a freedom to choose principle were to be employed, the supply side of this market would be complicated by the fact that users value data in different ways. An alternative to paying individuals for use of their data might be to require platforms to pay a "digital dividend" to the state or other organisation in recognition of user rights to their data. These alternatives all focus on individual rights of ownership in personal data and they do not so far address an individual's right to control the use of data derived from observing user behaviour.

5.4.2   Changing the public–private boundary in platform provision

Shifting the public/private boundary of platform provision in favour of either individual or collective control of online interactions might be achieved by interventions that favour alternative types of platform providers – those where the business model is not based upon commercial datafication. We consider three such alternatives, each with its own business model: subscription, public service provision, and clubs or voluntary communities. For the pursuit of alternative platform models to be worthwhile, they must capture the imagination of users and provide better support for public values associated with content and control than the existing datafication model. The three alternatives identified here are not all of the possibilities – other alternatives may emerge as the result of citizen, community or entrepreneurial imagination and innovation.

5.4.2.1   Subscription models

The first alternative is user subscription. As the term implies, user subscription platforms involve charging users for access. Subscription-supported platforms still might carry some advertising or engage in some commercial datafication, but, generally, they choose to be more constrained in the use of these practices in order to differentiate their offerings from "free" competing alternatives. An example of a subscription-based platform is Netflix which currently advertises only its own offerings. Subscription-based models imply the existence of desirable content whose value to users is greater than competing alternatives, including platforms using commercial datafication.

Digital news content is an illustration. Some news producers are creating "pay walls" for their content and requiring subscription before the full content of stories or other items is available to users. This could represent a possibility for "hollowing out" existing platform-based news aggregation practices because a growing share of *freely available* content that is aggregated as "news" and editorial content is contributed by individuals and organisations that are not paid for the costs of producing it. They do so with their own motives, which are not necessarily in accord with professional journalism standards. Thus, as the pay wall or subscription model grows, platform aggregation of news content may deviate further from traditional professional journalism standards. However, so far this growth is very gradual because of the small increases in the numbers of people willing to pay for news by such subscriptions. It is possible that public sentiment may change towards subscription-supported news media, but major shifts are not yet visible. At present, models for news and editorial products involving joint subscriber and advertising support are financially challenged by the content aggregating platforms (Pickard, 2020).[11] A different model or a marked change in user choice will be needed if this type of content is to be supported by subscription sponsorship on a large scale that is aligned with public values.

There are many other areas where subscription models may be relevant. Online education resources are delivered currently by a mix of freely available sites (often subsidised by universities) and subscription provision. Media streaming services offering films (Netflix again), music or other audiovisual services often are based upon subscription models. Many magazines which were published previously in print versions are now available online by subscription. Dating services provide a form of social media and, generally, are subscription based. These are all examples of desirable content that might compete for user attention with the dominant platform's services. Collectively, they add diversity and plurality to the online environment and, in this sense, they may improve public value outcomes. Such initiatives are accountable to their subscribers and, for this reason, they fall short of a general public value or online commons model of provision.

5.4.2.2   Public provision models

Public provision of platform services is a common model for e-government at national and more local levels. These services are enormously varied in scope and intent. Some are limited to providing citizens with an interface to government services, including fulfilling permitting, taxation and registration requirements. Others also aim to improve civic and democratic processes by encouraging social and political exchange among citizens, often most effectively at a local level. Perhaps the most emblematic of public provision models is the translation of public broadcasting into online public service media. This model is variously supported by voluntary contributions (e.g. the US), direct mandatory license or levy (e.g. the UK) or general taxation (e.g. Flemish Belgium). Radio frequency spectrum (a publicly controlled resource) was allocated in many countries to support broadcasters with public service obligations to inform, educate or entertain the public. The interpretation of public service obligations is often controversial, requiring ongoing elaboration in practice. Lurking behind, or directly at issue, in public service media obligations are concepts such as representativeness, inclusivity and an obligation to overcome political, ethnic, gender or religious bias. These issues were of concern with pre-internet limitations on content production and access that defined an era of scarcity, but they also arise in the era of content abundance since diverse content may be difficult to discover and representation biases difficult to detect and overcome.

The public service media model is under pressure in Western countries with differing historic mixes of public service and commercial media and this is exacerbated by platform news aggregation and subscription-based services. In the UK, for example, the BBC is facing a decline in television viewership, especially among younger people as they turn to online video forms of entertainment and news[12] – a pattern that varies throughout Europe and the US. Governed by a time-limited charter the BBC is required to provide news content, both general and specific, to UK regions and to named groups within the population with a remit to inform, educate and entertain all of its licence holders. The BBC, like other public service media, is continuously engaged in controversy about the exercise of its editorial duties that are claimed by various groups to be biased in favour of one or another viewpoint.[13] Controversy is endemic in the public service media model (Herzog et al., 2017; Lowe et al., 2017), making public service media vulnerable to political influence.

In order to demonstrate responsiveness to its public service mandate, the BBC's Data Insights Division undertakes data-led design, experimentation and audience analysis to provide assurances that its content is valued by audiences. One of its aims is to increase the number of "signed in" users on the iPlayer platform to acquire user data, especially on younger BBC platform users.[14] Although a form of surveilling users that is similar to commercial platform practices, it is done in support of a public service commitment to a common public interest and social solidarity. In addition, public service media in receipt of licence or state funding are often required to demonstrate the "value" they add to allay concerns that they unfairly compete with commercial enterprises with no public service mandate. In Europe, publicly supported media must not "distort trade and competition to an extent contrary to the common interest"[15] and a "public value test" is used to guide the appropriate scale and scope of public service media (Moe, 2010).[16] In view of the centrality of public service media content in a functioning democracy, "the vision of a truly public media—one that is genuinely accountable to and representative of publics and that scrutinizes elites rather than deferring to them—remains as relevant as ever" (Freedman, 2019: 214) in the digital platform era.

Public service media typically require public appropriations or voluntary donations to cover their costs. Because of these sources of funding, they are less willing and, in some cases, not permitted to bundle advertising in their offer or to monetise the monitoring of users.[17] In the face of pressure to adjust to the largest platforms, attention focuses on linking news and other forms of public service media to social and cultural goals, with an expectation that a service will serve the entire public. Another important example of this is online public healthcare systems which provide services allowing all patients to book appointments, request medication, seek advice or view other available services.

5.4.2.3   Clubs or voluntary community models

A third alternative approach to platform provision is "clubs" or voluntary communities. The concept of a club is drawn from neoclassical economic theory of behaviours that involve collective contribution and collective reward (Buchanan, 1965). One type of club is voluntary communities which creates a commons for delivery of content of interest to its members and perhaps is open to the public (Benkler, 2016; Ostrom, 1990). Early visions of collective efforts to establish virtual communities for information exchange were signalled by Mitchell (1999) and Rheingold (2000). Platforms such as GitHub or SourceForge which are designed to support open source software creation and the Wikipedia platform are contemporaneous examples. Club-based or voluntary community platforms have many business model configurations, but a central feature is that content is generated by a group of highly committed individuals. This means that the model for recovering costs needs to be aligned with the interests of the content generators.

These clubs or communities may be based on voluntary efforts (e.g. Wikinews) or on more complex business models including a mix of public and private support. The type of content created by clubs or voluntary communities generally reflects the interests of members who are likely to be significant sources of financial support and of audience. The general public may have access, though perhaps not complete access, to the content produced. The value of this model is that resources can be devoted to specialised production of content and its editing in service to its members who, in turn, have a voice in governance. Flexibility of governance means that members can experiment with a variety of ways to support the costs of generating and distributing content.

This model is illustrated by news about science (e.g. the UK's Royal Society or the American Association for the Advancement of Science (AAAS)). The Royal Society receives approximately half of its annual income from the UK government, while the AAAS receives over half of its income by publishing subscription-based scientific journals. Both produce news content and employ professional journalists and editors. Another example is Wikinews, an all-volunteer endeavour, which produces content subject to peer review by other members of the Wikimedia community. Some contributors may be professional journalists, but they are not paid by the Wikimedia Foundation, the owner of Wikinews. The residual costs of hosting and distributing Wikinews content are supported by voluntary donations. This means that, in an indirect way, Wikinews bears a considerable similarity to news publishers that have gone online and solicit user payments (e.g. *The Guardian*) rather than requiring subscription for complete access.

The principal shortcoming of the club or voluntary community approach is the specialised audience it creates, which is a central feature of its design. While this audience may, in fact, support the costs of professional journalism and editorship, this design does not address the need for diversity and plurality of news content. Another problem is that this model makes content production a by-product of other activities that pay the living costs of contributors, although the mission of the club can be financially supported by member contributions. If this fails, this alternative model may only be sustained by reverting to the advertiser-supported and/or subscription models. This model builds, however, on the opportunity to publish on the internet and offers an alternative to traditional models of professional journalism that are under pressure.

5.5   Conclusion

Self-regulation is, in principle, an effective means of assuring the alignment of public values and private interests without the cost and procedural complexity of external policy and regulation. A central contention by platform owners is that self-regulation is being practised to useful effect. To the extent that agreement can be reached about the public values that need to be embedded in platform operations, and practical means can be found for doing this embedding, platform owners appear willing to evolve their self-regulatory processes. A critical view of these claims, however, is that current arrangements do not provide an adequate amount of information to ascertain the extent of misalignment of platform practices with public values. The evolution of self-regulation is already leaving conspicuous gaps.

The three approaches to economic analysis we consider in this book differ on the presumption of the viability of competition as a means of achieving social benefit. They also differ on the extent to which it is necessary to take specific institutional rules and norms governing commercial platform operation into account. Institutional economic analysis suggests that strategic moves of the platforms should be investigated to identify potential negative outcomes and to assess the need for redress by creating opportunities for alternatives. In a critical political economy framing, dissatisfaction with self-regulation is common – it is viewed as obfuscation of true intent and behaviour. From this perspective, it is essential to devise means of resisting the progress of commercial datafication. Public service provision is a much more likely way to assure public values.

The neoclassical economic lens is more sanguine about outcomes due to its presumption that market competition will remedy untoward practices that do not sustain public values. State intervention generally is taken to be a last resort, legitimate only when market competition is demonstrably compromised by collusive behaviour or monopolisation.

Alternative platform provision models are more likely to accommodate and reflect public values, but they remain vulnerable to political influence or are self-limiting in their platform audience. Within the current institutional framework, these alternatives are unlikely to provide a foundation for a different path of future platform development which sets the stage for our examination of opportunities for external policy and regulation. A robust alternative platform provision model which can support costs in the face of challenges presented by dominant commercial platforms remains to be devised. This suggests that public values related to the creation of diverse and plural content capable of underpinning a flourishing public sphere are subject to indefinite erosion. In the case of digital content – and especially news content – its crucial role in sustaining democracy means that combinations of public service media and club-based or voluntary initiatives are likely to be the focus of future policy measures, and "public service journalism" based on obliging taxpayers or the platforms to contribute to production costs is being proposed (Pickard, 2020).[18]

The next chapter identifies external platform policy and regulatory interventions, and the economic arguments as to their usefulness and necessity.

Notes

1. See Harris Interactive (2019).
2. See Zuckerberg (2019).
3. See Google (n.d.).

4. E.g. Facebook's revenues rose 29 per cent in 2019 as compared to 2018 to USD 17.7 billion (Murphy, 2019).
5. See Algorithm Watch (2019).
6. We prefer user-contributed and user-interaction generated as descriptors as they better capture processes by which the data is acquired.
7. This includes revenues from major internet platforms, data brokers, credit card firms and healthcare data firms, see Foroohar (2019).
8. See Solid, MIT project at https://solid.mit.edu/ and The Hub of All Things, HAT Community Foundation project at https://www.hubofallthings.com/main/what-is-the-hat (accessed 3 January 2020).
9. See Lanier and Weyl (2018).
10. See, for example, NESTA (2019).
11. And see Bell and Owen (2017) and Newman et al. (2019).
12. See UK (2019c). The BBC's licence fee is assessed by households having a television and, because almost every household has a television, it is nearly universally applied.
13. See UK (2019f) for a review of public service broadcasting in the UK.
14. See Mari (2019).
15. See EC (2009: 21).
16. Public value tests were developed earlier by Moore (1995) in the context of public management in the US. See Barwise and Picard (2014) for an analysis which suggests that the whole of the media market may benefit from the presence of a strong public service media provider.
17. There are of course hybrid models such as operating a separate commercial division, providing a public service obligation to companies that rely on an advertiser-supported business model, and various types of programme sponsorship as in the case of the US model where advertising is institutional and confined to the beginning and end of programmes as opposed to advertising breaks.
18. And see UK (2019b).

6. Policy, regulation and alternative platform provision

6.1 Introduction

When platform self-regulation falls short of expectations, can external policy or regulatory interventions achieve better outcomes? And, what are the limits or unintended effects of such interventions? All three of the approaches to economic analysis are consistent in their position that external policy and regulation involves more than economic considerations. Neoclassical economics ordinarily does not advocate intervening in market operation because market competition is seen as the most effective regulator of behaviour. However there are exceptions. A company which has market power can dictate price or impose requirements on suppliers or customers without an effective competitive response. This response may come from existing rivals or the potential entry of new competitors. Intervention is justified when the competitive response is inadequate (e.g. too small) or impossible (e.g. blocked by collusion). In addition to measures to address market power, neoclassical economics acknowledges that democratic societies may make political decisions to impose regulations or to establish mechanisms for producing public or merit goods for which market incentives are inadequate. When they do, these economists can provide some indication of the consequences of these decisions by assessing the effects on income or wealth distribution.

Institutional economic analysis acknowledges greater scope for uneven relational power. It is particularly concerned with norms, rules and standards that might extend or reinforce the ability of some actors to influence or coerce the actions of others. A critical political economy analysis understands coercive power as ubiquitous in a capitalist economy and seeks means to constrain or contest the exercise of this power. In both these analytical traditions, fundamental structural change may be considered aimed at changing market behaviour or at moving the boundary between public and private provision of platform services. This chapter reviews external policy and regulatory measures, including enlarging opportunities for alternative platform provision.

6.2   Rationales for regulatory intervention

Uncertainty is involved in predicting outcomes of regulatory measures to limit the digital platforms' power. Uncertainty introduces a political element into decisions about whether to modify or introduce new regulatory provisions (Marsden, 2018b). For example, it can be argued that departures from platform self-regulation may succeed only in constraining the behaviour of a few "bad actors" while reinforcing and extending the interests of incumbent companies (Marsden, 2017, 2018a). The centrality which neoclassical economics places on the goals of competition and individual freedom of choice means that external regulation should only be undertaken when these goals are compromised. The form of regulation should demonstrably lead to better outcomes than competition based upon freedom of choice of both suppliers and customers. As noted earlier, neoclassical analysis is also capable of analysing the impact of political decisions to regulate with the aim of identifying unintended or untoward outcomes such as the creation of new means for limiting competition.

As we have also seen, the institutional economics lens offers a broader view of platform performance. It accords public values, beyond competition and individual freedom of choice, a standing in the assessment of their performance. From an institutional perspective, regulatory interventions should lead not only to a healthier and more robust commercial market, but also to a public sphere in which citizens can engage in ways that are consistent with public values. Short of replacing capitalism, both the institutional economic and the critical political economy frameworks focus attention on the boundary between public and private values. Could public values be better served by regulation that somewhat, or dramatically, reduces the exercise of power by the platform enterprises?

The results of the application of these economic lenses feed into political contests over appropriate regulatory mandates for external actors. Cohen (2016: 369) argues that "regulatory institutions are most usefully understood as moves within a larger struggle to chart a new direction for the regulatory state in the era of informational capitalism". What configuration of institutions is best placed to encourage norms and rules consistent with societal expectations and to bring them into practice? There already is a mix of legislation and regulators whose mandates encompass some aspects of the platforms' activities. This means there are challenges in coordinating regulatory actions within countries as well as interjurisdictional coordination problems when these actions diverge.

In the US – the home of many of the largest digital platforms – the government, until recently, has taken a technophilic approach to platform regulation, treating these companies as a vital source of wealth creation and economic growth. By contrast, from the 1870s to the early 1900s, and again during the Great Depression (1929–1939) after Franklin Roosevelt was elected as President in 1933, the US government was more likely to treat the monopolistic practices of large companies with suspicion. In industries such as rail transport and oil and steel production, this led to enactment and enforcement of antitrust laws aiming to curtail their market power. Exclusive dealing, price discrimination, refusals to allow access to what were deemed essential facilities, tying or bundling of products and predatory pricing, as well as refusals to make patents available, were addressed as abuses of dominant position (Wu, 2018). Concerns about "bigness" continued into the 1970s, when IBM was forced to unbundle its software from its hardware; in the 1980s, when AT&T was broken up; and in the 1990s, when Microsoft settled a case agreeing to curtail some of its anticompetitive practices.

By the end of the 1980s, however, the argument was that the threat of potential competition is generally substantial enough to discipline the actions of large companies, although explicit collusion such as price fixing was still subject to intervention. In addition, the size of foreign companies had become a concern and arguments were made that potential rivalry from abroad could act as a further limit to anticompetitive practices. Advancing international competitiveness became a justification for preserving large enterprises. For all these reasons, there was little interest in taking actions to prevent digital platforms from becoming very large companies. In Europe, there was greater attention to whether the dominant (mostly American-owned) platforms were creating barriers to the entry or growth of European firms. Competition authorities investigated, for example, the platforms' ability to control personal data, privilege the ranking of information, or aggregate datasets in ways that were disadvantaging competitors. Despite such investigations, however, proposed mergers and acquisitions, often of firms with European origins, by the leading platforms, were rarely prevented (Kadar and Bogdan, 2017).

Sparked by disquiet about the behaviour of the largest platforms, and especially Facebook's and Google's continuing inability to operate in a way that is deemed to safeguard public values, there are signs of a move to regulate "by outrage"[1] in Europe and in the US. In the UK, the large social media platforms have been labelled, for example, as "digital gangsters" and charged with engaging in "evil" practices. In the US, Facebook has been called a "disinformation-for-profit machine" and social media has been described as a "corrupt system".[2] There is a perception among policy makers that the operations of digital platforms do have negative consequences sufficient to provoke a response. This is especially so as evidence accumulates that individuals or targeted groups are vulnerable to misinformation, to being nudged and manipulated, or to being harmed due to the absence of reasonable levels of critical digital literacy. Governments are considering or enacting measures that are expected to influence the platforms' operations and, in some case, to restructure markets. Policy makers are seeking new norms and rules to limit illegal speech, reduce harmful information or control "bad" online behaviour (especially when it harms children) in the Western democracies. However, they also acknowledge obligations to ensure that the right to freedom of expression is not abridged as a result of new forms of platform regulation.[3] The question is which norms, rules and standards should be followed?

The regulatory measures that are introduced which bear on the platforms' operations may operate within current norms and rules governing the right of corporate actors to operate commercial business models in pursuit of their lucrative datafication strategies. Additionally, policy measures may be structural in the sense that they seek to alter the platform market or to give higher priority to collective non-commercial models of platform service provision. In both cases, the aim is to stem actual or possible harms associated with commercial datafication.

We start with a discussion of measures aimed principally at modifying the operation of basic platform business models and making these operating models more transparent and accountable to the public. This is followed by a discussion of measures aimed at changing more fundamentally the markets in which the platforms operate, that is, structural remedies. The discussion in this chapter leaves open the question of whether decisions to introduce new norms and rules for platforms should be implemented through state/corporate co-regulation, unilateral state initiative, or by establishing one or more independent institutions. This is because institutional set-ups vary substantially across societies.

6.3   Regulating platform operations and markets

Four areas are prominent among the measures aimed at creating incentives for the platforms to modify their operations: the introduction of government mandated responsibilities; data and privacy protection measures; the use of codes of practice; and measures to strengthen skills and training policies.

6.3.1   Government mandated obligations

Discussions about new forms of platform regulation aimed at holding the digital platforms to account often focus on how such accountability is to be achieved. Operational incentives to ensure that the platforms sustain public values may be introduced through a statutory "duty of care" (a moral or legal obligation to ensure the safety or well-being of others) or by designating platforms as "information fiduciaries" (legally enforceable promises to act in a certain way).[4] Such measures are likely to be contested by those arguing that platforms should be acting as "responsible guardians" of public values (Gillespie, 2018) under their current self-regulation strategies. An external policy and regulatory role for government would not be a novelty since traditional digital content companies have been subject to regulation historically, while newspapers were held to self-governance and legal standards. However, the platforms' business models and practices bring into stark relief the conflicts among values such as privacy protection and freedom of expression. This is both because of the global scale of their operation and because of the opaque nature of their uses of the technologies of datafication.

Insofar as the platforms engage in *media-like* editorial processes as a result of their content moderation practices, what measures should apply when new forms of oversight are introduced to protect freedom of expression? The platforms argue that those responsible for illegal content should be the focus of law enforcement and stand ready to assist such enforcement. Nonetheless, there is heightened concern about the spread of illegal content on platforms. The overriding consideration has been acceptance of the principle that platforms are conduits rather than publishers of information. They mainly are exempt from liability for illegal content when serving as a "mere conduit" or caching content in which they are not involved in production and do not modify. This is enacted in law. For example, platforms may not be given a "general obligation to monitor online expression" under the European Union's e-Commerce Directive 2000.[5]

Platforms providing "information society services" (at the time of writing) were, however, expected to implement procedures for removing and disabling access to illegal information on a voluntary basis. When a platform becomes aware of illegal or infringing content or activity, it must act "expeditiously" to remove or disable it. In 2018, a European Commission recommendation called for *mandatory* platform compliance regarding illegal content. A new European Digital Services Act is under discussion at this writing which may introduce provisions that make platforms, or their senior executives, liable for illegal content.[6]

Whether platforms operating in Europe should retain their conditional liability is being reconsidered in the light of claims of harmful content, e.g. content that is harmful to children, young people and adults. Examples of harmful content include depictions of self-harm, advocacy of suicide, false claims of immunisation harms and content aimed at denigrating class, gender or ethnic origin. Despite the many passionate advocates for limiting such harmful content, limits on expression are very difficult to implement without jeopardising the fundamental right to freedom of expression.

In the US, the Constitution protects speech rights, limiting the extent to which platforms (or publishers) can be deemed liable for the content they host. Thus, even if some expression might be deemed harmful, it is unlikely that platforms can be held liable for hosting content in the US. This is reinforced in legislation. Platform responsibility for content is determined by the Communications Decency Act of 1996 which states that "no provider or user of an interactive computer service shall be treated as the publisher or speaker of any information provided by another information content provider".[7] This provision confers immunity on digital platforms including those publishing third party content. Exceptions concern criminal content (expanded after 9/11) and content protected by intellectual property rights which are subject to legal take-down requests. In Europe, speech rights are not subject to such absolute protection, e.g. Germany's restrictions on expressions regarding unconstitutional political organisations such as the Communist Party or National Socialism. New regulations limiting content deemed harmful are being considered. When such proposals introduce new platform responsibilities or liabilities for content, they are resisted by arguments that this increases the likelihood that they will err on the side of caution and abridge the right to freedom of expression.

6.3.2 Data and privacy protection

Datafication processes, including the uses and consequences of algorithms, need to be transparent if the platforms are to be held to public account. Concerns about data protection and individual privacy have been raised since the earliest phases of the computerisation of networks (as discussed in Chapter 1).

In Europe, regulatory constraints on the platforms' behaviour in this area were introduced through legislation such as the Directive on Privacy and Electronic Communications which was agreed in 2002 (and updated in 2013). A General Data Protection Regulation (GDPR) came into force in 2018 which applies to the processing of personal data including by automated means.[8] Rights are conferred on platform users to be informed of how their data is used, to access this data and, as appropriate, to seek rectification of errors. Rights are also granted for erasure or "to be forgotten", to restrict data processing and to object to automated decision making and profiling. The platforms are expected to operate lawfully, fairly and transparently and to hold or use data only for limited and specific purposes.

In the US there is no overarching federal legislation, and privacy and data protections are provided to online users through multiple provisions. This is despite an early Federal Trade Commission (FTC) observation that "industry's efforts to encourage voluntary adoption of the most basic fair information practices have fallen short of what is needed to protect consumers".[9] The Federal Trade Commission Act contains provisions enabling the FTC to protect consumers against unfair or deceptive business practices and these are augmented by various state laws. Among the most recent is the California Consumer Privacy Act which includes provisions such as a right to request access to and the deletion of personal information and to opt out of having personal information sold to third parties.[10]

The principle underlying these actions is to give individuals the right to opt out of granting platforms permission to collect, use or sell their personal data or data which is linked to them. With respect to personal data, the individual is expected to know what information is being collected and how it may be sold or disclosed and to be fully (or reasonably) informed before consent is given – an approach which has been characterised as a "pathology of consent" and as a "broken" system that puts the onus on the individual to know the risks they may encounter when they go online (Richards and Hartzog, 2019).[11] These provisions concerning personal data have little direct bearing on how the platforms handle non-personal data (e.g. observation of user behaviour) which remains largely subject to their discretion as set out in terms of service agreements. Some argue that protecting individual privacy and achieving greater equity requires that platform users should have a right to opt in to the collection or sale of data about them; others argue that there should be areas where decisions based on, or informed by, data analytics and algorithms are not permitted.

In Western regions and countries, a variety of individual rights or risk-based data protection measures and tools has been introduced, including establishing data protection commissions with capacities to levy fines when breaches to personal data protection and privacy legislation occur (Bennett and Raab, 2018). However, as the platforms amass user-related data and the IoT develops, generating even greater amounts of data, much of it non-personal, the risks of privacy intrusions and other harms are increasing. These are linked to wider issues of surveillance and the ethical use of data collection techniques (as discussed in Chapters 4 and 5). The absence of transparency in the way digital platforms engage in datafication is resulting in efforts to update legislation, but these efforts are challenged by rapid technical progress in datafication practices.

6.3.3   Codes of practice

If new harm-specific legal obligations are introduced or steps are taken to introduce a duty of care or a fiduciary responsibility, the next step is to make these operational in platform codes of practice. The expectation typically is that such codes should result in the same level of protection for consumers and citizens online as they are entitled to offline. Such codes may include requirements for platform accountability and transparency (including revelations about the use of algorithms) and for the evidence needed to enforce them (Flew et al., 2019). They may also govern expectations for content diversity and plurality and the treatment of illegal and harmful content. Requirements for the platforms to make use of technical tools such as content flagging may be aimed at supporting audience abilities to discriminate between accurate and misleading digital information sources.

Technical standards may also be covered by such codes. Mandatory open standards for interfaces between platforms, for instance, might be introduced to enable users to switch platforms more easily (see Chapter 5). The platforms may then have incentives to start offering differentiated levels of data and privacy protection, paid-for services without advertised content, or to give users more control over the content they see. However, this regulatory move assumes clarity about what user data should be portable and what "user data" means in practice. Should data portability mean that users can take their timelines from Facebook and mount them on another platform? This might assume a transfer of the address identities of those who have commented on the user's timeline (hence, moving data about them that Facebook owns, but the user requesting the data does not). It is also unclear whether person-identified data files of observational data exist as they may be unnecessary for many types of data analytics. These kinds of measures aim to give platform users greater control over their online interactions, and, specifically, their personal data, but implementation in a rapidly changing technical landscape is likely to prove very challenging.

Codes of practice are also relevant for the standards associated with ensuring that algorithms used to rank, select and filter information are "neutral" in the sense that they do not favour their own and their affiliates' services and content over others. Any requirement for algorithm "neutrality" assumes a high level of transparency, and this is problematic given the machine learning techniques employed to create predictive systems (see Chapter 4). Some will argue that these algorithms are the result of innovative private investment and should remain proprietary to preserve innovation incentives. Further complicating this is that algorithms often rely upon training data which, as observed in Chapter 4, incorporates unintended and unknown biases.

A standard embodied by a code of practice may also address the ethical issues related to platform behaviour and, specifically, the concerns raised by AI and machine learning (see Chapter 5). One such standard is that a platform's AI-driven systems used to make or support decisions should be designed to give intelligible reasons; to meet an "explainability" standard. What rules define explainability, however, remain to be determined, either over an extended period of judicial interpretation or through yet-to-be-articulated legislative efforts. Alternatively, a standard might require that algorithms be "contestable" in the sense that their adherence to stated principles is subject to contest by a reviewing body, such as an algorithm ombudsperson, again suggesting a protracted accumulation of precedent.

6.3.4 Skills and training policies

The use of AI-related systems to augment (or displace) human performance is growing, although there is uncertainty about the rate at which changes in the workplace and in labour relations will occur in the digital economy (as discussed in Chapter 4). It is being recognised by policy makers, however, that both children and adults will need new skills. The response has been to acknowledge the need for lifelong training and retraining. There have been calls for apprenticeships and for the redesign of all levels of training and education. For the most part, efforts to develop a workforce that is adequately trained have relied on encouragement by governments to increase investment in cooperation or partnership with industry and universities, with limited efforts to overcome gender and racial biases in training.

Skills and training initiatives perform a platform governance function in relation to public values insofar as they inculcate understanding of the potential risks and harms associated with online interaction. Internet user confusion among adults and young people is widespread about how the data they provide, or that is gathered based on observing them, is being used.[12] Evidence of declining trust in online sources of information,[13] and of the spread of illegal or harmful information, is calling attention to the need for regulatory interventions to aim at strengthening the development of critical literacy skills among all those who engage with platforms. Policy interventions in this area include investment in tools and knowledge and the introduction of incentives to encourage adults and children to acquire skills to navigate online spaces effectively and safely. This involves encouraging (or requiring) critical literacy training to help people to make better assessments of the accuracy of online information (Livingstone and Wang, 2011).

The question is who should bear the costs of addressing the need for enhanced skills and training? Digital literacy training historically has received limited state funding and the commitment of the private sector has been uneven. In the UK, the regulator, Ofcom, has had a mandate to address media literacy since 2003, and the European Union's Audiovisual Media Services Directive (AVMSD) has required reporting on progress in this area. In practice, however, support for media or digital literacy has been neglected. An updated AVMSD in 2018 now defines media literacy (beyond learning about tools) to include critical thinking skills,[14] and this is expected to encourage new measures in this area.[15]

It is essential, however, to clarify where responsibility lies to implement media literacy measures. In the UK, legislation is pending following consultation on an *Online Harms* White Paper. This is likely to include the ambition of developing a new online media literacy strategy.[16] In the US, a Digital Citizenship and Media Literacy Bill was under consideration by Congress at the time of writing which would provide grants to state education agencies.[17] On their own, however, measures in this area cannot be expected to fully address the problems created by a ubiquitous online digital environment. They need to be accompanied by measures to ensure that the platforms' operations uphold public values and foster behaviour consistent with those values.

6.4 Structural intervention and alternative platform provision

Policy interventions can change platform behaviour through modifications in the structure of platform companies or in the markets in which they operate. These interventions may employ competition policy and antitrust provisions or seek to create or enhance the prospects for alternative business models (as discussed in Chapter 5) that aim to shift the boundary between public and private platform provision.

6.4.1 Competition policy and antitrust

Neoclassical economic analysis, as we have seen in earlier chapters, places social and political concerns about platform dominance outside the application of competition policy and its antitrust provisions. Conventionally, competition policy and antitrust provisions aim at ensuring that competition is fair and advances consumer welfare. The latter typically is understood narrowly as the benefit achieved by consumers in being free to choose which goods and services to acquire. Is there evidence of harmful anticompetitive platform behaviour that raises prices or curtails choice?

In current practice, the answer depends on determinations about effects within a construct called the "relevant market". This is defined by whether there are existing, or readily created, substitutes for products or services offered by a dominant or merged company that would limit a platform's ability to raise prices or change conditions of service. As Bork and Sidak write, "if Google or any search provider caters too much to advertisers – by, for instance, ranking natural search results according to payments from advertisers – it risks losing search engine users who are not finding the results they prefer" (Bork and Sidak, 2012: 4).

A broader interpretation of consumer welfare may, however, provide a means of influencing platform development in ways that align better with public values. For example, in the platform market as explained in Chapter 3, there is a relative scarcity of attention which affects the resale value of platform users' attention and the price that can be charged to advertisers. In the antitrust context, there is discussion about the appropriate standard for assessing platform market competitiveness using measures to assess dominance and its impact on consumers who are unprotected from "non-consensual and entirely uncompensated transfers of attentional resources" (Wu, 2019: 805). Short of a moratorium on all platform mergers and acquisitions, it is also suggested in the US context that the existing consumer welfare test can be applied effectively if there is the political will to launch antitrust investigations (Khan, 2017). In Europe, the narrow consumer welfare test has also served as the basis for determining whether the platforms are engaged in anticompetitive behaviour. This may be changing, however, as competition authorities begin to consider non-price criteria such as harms to privacy, quality of service and innovation (Just, 2018).[18]

If platforms are found to be engaged in anticompetitive behaviour, competition authorities have several remedies: they can fail to approve a proposed merger or acquisition; they can impose requirements for the internal structural separation of lines of business; and/or they can require the break-up of a company. It is unclear how far departures from narrow interpretations of market failure will be taken. This is because competition authorities are likely to argue that the benefits of the digital platforms should not be lost as a result of inappropriate or disproportionate interventions.

In the face of increasing concern about the implications of the large platforms' dominance, investigations under competition law provisions have been opened by the European Union's Competition Directorate and, in the US, by the FTC and the Department of Justice. These focus on potentially anticompetitive practices in relation to the use of sensitive data and on whether the platforms are reducing competition, stifling innovation or harming consumers.[19] Broader interpretations of anticompetitive behaviour may provide a rationale for intervention as a result of these investigations in the form of the break-up option. There is, nevertheless, likely to be a continuing lack of clarity about the extent and nature of the interventions that can be justified under existing law to create "healthier" platform industry market structures. Diminishing the scale of the largest digital platform operations by breaking them up to create smaller companies faces several problems.

If some platforms are, in fact, "natural" monopolies due to the existing configuration of laws and norms, breaking them up would create inefficiencies that need to be weighed against the claims about the harms of large size. The resulting smaller company structure could also create challenges for regulators to implement other measures aimed at increasing the transparency and accountability of, for example, their datafication practices. These would need to be applicable to a larger number of business operations and be monitored effectively. In addition, because much data accumulates as a result of users who share their information with others, a larger number of smaller companies also might result in excessive data sharing and further risks to privacy (Acemoglu et al., 2019).

6.4.2   Alternative platform provision

The models for alternative platform provision discussed in Chapter 5 suggest consideration of the competitive alternatives to ever-greater reliance on the commercial datafication-driven platform model. Enlarging the space in which these alternatives might flourish and place greater competitive pressure on the dominant platforms can be achieved by specific interventions to change the norms, rules and standards for platform provision. Some would be fairly dramatic in their impact on the boundary between public and private supply of online service. We consider the three models for alternative platform provision in turn.

Uncertainty about the viability of the subscription model exists because platforms are actively engaged in acquiring or creating content that might otherwise have been used as the basis for subscription-based access. For example, news aggregation and contributions to news and editorial sources weaken the subscription-based model in journalistic content. In principle, subscription-based services would be in a stronger position if they were able to bargain collectively with platform providers over such contributions. Creating this collective position, however, is likely to be problematic in jurisdictions which discourage cartels. A possible alternative is to create a commission independent of content providers and platform owners tasked with the negotiation and mediation of agreements. To do this, content providers would have to be in a stronger position. In some content areas, strengthening intellectual property protection by expanding its scope or better defining infringement penalties might achieve this.

In journalism content, one means to strengthen publisher and broadcaster positions would be to impose public interest obligations on platform providers. Such a reform might begin by tying this obligation to the reputation of platforms; if this is insufficient, then by levying specific fees to discharge the obligation. Unlike the regulation of services which use the radio spectrum, there is no licensing requirement for a platform company, so it might also be necessary to translate this obligation into a tax credit offsetting those taxes that platform companies are obligated to pay on their revenues in different jurisdictions.

International tax law permits platform companies to allocate their costs and revenues across jurisdictional boundaries and, thus, to reduce taxation on profits in any particular jurisdiction if they are not legally physically established in a country. Revisions to tax law are being considered by the OECD and G20 countries in the wake of claims that the largest platforms are using existing law to minimise their tax payments in the countries in which they are active. To realign the location of taxable profits with the location of the platforms' economic activity and value creation, proposed measures would change the definition of "permanent establishment" and reallocate some profits and taxing rights to countries in which platform usage is growing.[20]

Proposals under discussion are intended to pre-empt countries from making unilateral changes. Individual countries are nonetheless considering national level "digital taxes" or "dividends" to sustain news publishing and to encourage content diversity and plurality. Whether justified using a neoclassical public or merit goods argument or by invoking public values arguments linked to citizens' rights, the risk is that new taxation arrangements will be treated as a nontariff trade barrier which could result in unintended consequences or in incentives for the digital platforms to reduce or withdraw services from a country.

The process of modifying tax provisions opens further opportunities to consider the public service provision model. Public service media have public service obligations that are associated with public values. Experience with its governance offers an initial benchmark or standard for considering the level and variety of requirements that could be addressed by a public service obligation on platform companies. There is always some risk, however, that entangling public provision with the state will diminish the independence of public service media unless there are clear provisions to maintain their autonomy. Beyond this, governments may consider funding the provision of infrastructural elements like data trusts or public infrastructure (complementing the large investments in infrastructure made in support of research conducted in universities and government). Ultimately, public service platform provision requires support from some form of taxation and the scope and extent of this provision is an important policy issue for national governments in both the global North and South.

"Commons based" or club and voluntary provision of platform services might encourage greater moral legitimacy and greater equity (Benkler and Nissenbaum, 2006). However, Hepp (2016) cautions that, even in collaborative open technology "maker" communities, achieving better designs that intensify commercial digital production and consumption may be the goal. In addition, this alternative business model is still likely to depend on AI-related innovations to attract users. Although the algorithms they use might be programmed in different ways and they might adhere to different community-agreed practices that safeguard user data, they confront the challenge of achieving "explainability" and the potential for unknown biases to occur. Norms and rules for content management are needed to respond to their members' behaviour, and controversies in judgements about harmful content are also likely to be experienced. The risk of censorship remains.

All of these alternative approaches do, nevertheless, benefit from their potential to distance themselves from state and corporate interests in commercial datafication that locate the boundary of supply resolutely on the side of the private market.

6.5   Conclusion

Which individuals or groups should bear responsibility for making decisions about the acceptability of platform operations and whether to take down content or to leave it up? This question will continue to be contested. How algorithms can, or should, be used to trigger attention to harmful or illegal content will also remain controversial in efforts to calibrate standards to local, national or global expectations.

To avoid "regulation by outrage", consensus is needed about what constitutes "good" and "bad" platform behaviour. In a period when any such consensus – especially globally – is increasingly difficult to achieve, the dilemma is how to ensure that harms associated with platform operations are not used as justification for giving (Western) governments powers that infringe on citizens' rights to freedom of expression and privacy. There is always a danger of overreach by the state and, in the case of social media, of a failure to provide independent mechanisms for discriminating between content and activities that fall within a (contested) sphere of acceptable public discourse and debate and that which is harmful. Organisations which do not seek to engage in commercial datafication practices are not immune to these risks, but they may have a greater likelihood of self-governance in support of public values.

Regulatory measures which seek to treat the platforms as "editors" risk replacing one problematic set of platform practices with state or independently imposed rules and norms that put citizen rights at risk if they result in unaccountable censorship. In seeking to counter the disproportionate and unaccountable responsibility that rests with the commercial platforms, policy and regulatory responses inadvertently, or by design, can infringe on the public values. Yet it is these values that are necessary to sustain digital spaces where a reasonably high-quality public discourse and inclusive participation can flourish.

In Chapter 8 we return to these issues. Before doing so, we consider some of the implications of the rise of platforms for those who are located beyond the Western democracies.

Notes

1. See doteveryone (2018: 20).
2. See UK (2019a, 2019g); and for the US, Culliford (2019) citing Senator E. Warren, and Paul (2019) citing Senator S. Brown.
3. For platform regulatory inquiries in the UK, US and other Western countries, see Morton et al. (2019: 109–119) and Puppis and Winseck (2019).
4. See UK (2019g) and Balkin (2016).
5. See EC (2000).
6. See EC (2018b).
7. See US (1996: 138).
8. See EC (2002, 2016).
9. See US (1998b: ii–iii).
10. See US (2018a).
11. And see UK (2019h).
12. See CIGI (2019).
13. See UK (2018a).
14. See EC (2010: Art. 33) and EC (2018a).
15. See Livingstone et al. (2017).
16. See UK (2019d).
17. See US (2019a).
18. And see Crémer et al. (2019: 3) for a report prepared for the European Commission stating that "even where consumer harm cannot be precisely

measured, strategies employed by dominant platforms aimed at reducing the competitive pressure they face should be forbidden in the absence of clearly documented consumer welfare gains".

19. See, e.g., EC (2019a) and US (2019b).
20. See OECD (2019).

7. Global perspectives

7.1 Introduction

In government, industry and some citizen discussions, a frequent assumption is the "digital imperative". This is the necessity of ever-expanding connectivity and universal adoption of a proliferating array of online devices. In short, it is assumed that to thrive in the 21st century, everyone must be connected and able to access digital platform services. This assumption is amplified in discussions of the "Fourth Industrial Revolution" (Schwab, 2017). These discussions envisage transformative change in industrial technology, following previous surges of change resulting from mechanisation, electrification of mass production and information technology-based automation. According to the World Economic Forum, the Fourth Industrial Revolution involves the pervasive application of AI, robotics and bioengineering. The digital imperative promises dismal prospects for those left behind, with proponents claiming that countries, and especially cities, that fail to participate in the newest revolution will diminish prospects for their workers and citizens.

This chapter examines the claims made for the necessity for countries, often located in the global South, to "catch up" with and prioritise digital platform and Fourth Industrial Revolution developments. We begin by examining the gaps between countries and the presumed universal movement to catch up and converge with leaders. This discussion raises questions about the limits of a digital economy contribution to economic growth and development. Choosing a different development path is then considered, followed by an examination of China's experience with digitalisation and platforms.

7.2 Being left behind

Virtually everything said about digital platforms in the previous chapters assumes relatively low or no barriers to becoming a platform user and few, if any, barriers to platforms offering their services. In so doing, we have omitted a large share of the world's population. We also neglect that part of the international growth strategy of platforms that seeks to overcome barriers such as problems in connectivity, users' knowledge of information and communication technology, or the presence of less permissive legal and regulatory frameworks in the global South.

The social and technological changes accompanying the information and communication technology revolution (in the WEF framework, the Third Industrial Revolution) were concentrated in the countries of the global North. Differences in patterns of digital development have created large divides between global North countries and the countries of the global South (which are also present *within* countries and regions of the global North). Lower investments in digital technology, in the connectivity infrastructure and in skills to use these technologies are all characteristic of the countries of the global South. The digital imperative predicts that the result of these lower investments will be slower rates of productivity and output growth and lower levels of prosperity in countries of the global South. Many countries in the global South, however, have markedly higher *rates* of output growth (e.g. measured in GDP) than countries in the global North. Although this is a hopeful sign, it is also the consequence of arithmetic in that it is harder to generate a high rate of growth on a higher base than a lower one. The differences in the *levels* of GDP between global North and South countries remain large despite some narrowing in recent years.

A corollary of the "digital imperative" is that "catching up" with the pattern of development experienced in the global North will have a direct and beneficial impact on society. These expectations are held notwithstanding the fact that around one per cent of the world's population holds over 35 per cent of global private wealth, more than the bottom 95 per cent combined. One in nine people go to bed hungry every night and one in ten earns less than USD 2 a day.[1] When it is argued that a catch-up process is possible it is therefore relevant to estimate the time required to close the gaps. It is also important to assess the effects of catching up. For example, a structural shift towards higher skilled labour required in the digital economy risks displacing lower-income jobs and thus increasing inequalities in the distribution of income. Even if (improbably) domestic rather than foreign companies dominate, similar patterns of development with the global North will concentrate returns to technology investment and trade in a few companies (Atkinson, 2008). It stretches credibility to assume that the greater accessibility of commercial digital platforms would be sufficient to address these imbalances even if the direction of causality did go directly from digital uptake to economic and societal benefit. Yet countries, regions, communities and individuals who lack connectivity (or face very expensive connectivity) and relevant digital skills are expected to acquire them so that they can close the gaps and catch up with the leaders.

The relationship between inequality and digital connectivity investment is complicated because it depends upon multiple factors. Over the past 25 years approximately the same number of countries has experienced an increase as has seen a reduction in inequality. Bauer (2018: 340) finds that those experiencing a decline in inequality had high and continuously improving levels of skills and education, enacted programmes to help workers and businesses adapt to new market conditions, and had policies directly aimed at mitigating income inequality. Studies that attempt to test the causal relation between changes in the use of digital technologies (e.g. use of the internet) and economic growth or productivity gains at the level of individual countries are inconclusive. Further decomposing growth processes within countries reveals that, even while the poorest are often becoming less poor, the gap between the rich and the poor is widening. This is because economic growth, in which digital technologies play a significant role, is disproportionally benefiting higher income groups and countries.[2]

"Digital divides" or inequalities, gaps in the rate of take-up and uneven experience of digital technologies by region, gender, ethnic group, socio-economic income level and other dimensions, are the subject of numerous consultant, United Nations and World Bank reports. Invariably, they pronounce that the next generation of technology can be expected to help enable catching up with leading countries. By contrast, academic research emphasises that digital inequalities arise in specific social and temporal contexts and cannot be understood simply by focusing on the diffusion or take-up of hardware or software (Helsper, 2017; Robinson et al., 2015; van Dijk, 2013).

Attempts to measure the extent of gaps and promising or discouraging trends in their closure reflect the presumption that digital technology developments should proceed throughout the world in a similar fashion; what economists call "technological convergence". To the extent that a country or region is not closing the gap, i.e. converging with richer countries, the claim is that they are being "left behind" and are therefore likely to suffer the consequences of lower growth and productivity improvement.[3] Criticism of this approach should not be taken to mean digital technology developments are inconsequential or irrelevant for countries in the global South. This critique does, however, open opportunities to investigate how these developments may combine with other patterns of change to produce or diminish positive outcomes.

7.3 Doing digitalisation and datafication differently

There are alternatives to the model of "technological convergence" in which all countries and all areas within countries run the same race towards higher levels of digital technology investment and user engagement with digital platforms. These include accommodations and adjustments to some elements of the technology-led model and bolder efforts in line with, or departing from, patterns of development in the global North. These alternatives can be distinguished by the extent to which they attempt to adapt and accommodate, or to overcome the centrality of, technology as a developmental force, and by differences in the digitisation and datafication processes they envisage and attempt to implement.[4]

7.3.1 Mitigating measures

Mitigating measures are those which seek to moderate or lessen the impact of problems created by digital divides including inter-country and intra-country differences such as rural–urban divides, regional disparities and gender or racial inequalities within countries. Although such measures often place a high priority on catching up, they often involve adjusting familiar, or adding new, elements to compensate – never easily – for differences between the contexts in the global South and North. The principal areas addressed are investments in connectivity and internet-based enterprises, education and skills, and accommodation to local contexts and regulations. Each is discussed in turn below although they are often interrelated.

7.3.1.1   *Investing in connectivity*

Digitisation is a major investment. Where there are clear opportunities for returns on digital investment in the global North, there is often an adequate available supply of private investment. Nonetheless, in the global North, there are parts of society that are not able or willing to provide the returns needed for private investment. Where this is true, measures are often undertaken to extend connectivity to those with no or modest ability to pay for digital services or to invest in training and education in digital technology use. A key measure is universal service obligations for communication or internet access service companies that, because of regulatory requirements, are only granted access to markets contingent on providing service to almost everyone. In the global North, training and education have received varying, but compared with the global South, significant, public support through incorporation in school curricula and provision as a social service for job seekers, the elderly and others.

In many areas of the global South, the scale of investment required to provide universal service is well beyond the means of either government or the surpluses available to companies. This leads to differences in the understanding of what connectivity might mean in the global South. Instead of universal individual access to internet connectivity, for example, a more feasible goal and a frequent target of such measures is neighbourhood or village connectivity to shared facilities such as internet cafes, libraries or schools.

Understandings of connectivity are also influenced by differences in the development of services in the global South where wireless communication has developed more rapidly than fixed line communication. For example, in many parts of Africa when extending fixed line communication was beyond the means of private or public investment, mobile service networks have been able to generate a return on investments. However, the mobile communication networks that are built often lack the capacity to service smartphones or to provide full access to platforms. A response to the bottlenecks created by lower connectivity investment has been to modify platform services. Private initiatives in a number of countries in the global South offer "basic" internet services as a substitute for direct access to the internet (which incurs high data charges) under "zero rating" initiatives (Hoskins, 2019; Romanosky and Chetty, 2018). While this helps platform owners with their global expansion strategy, it also reinforces their market control by raising barriers to local competitors including fledgling (local) platform companies.

7.3.1.2 Investing in education and skills

Mitigating measures involving training and education in the global South bear some similarities to those undertaken in the global North with an emphasis on STEM subjects and efforts to provide related training to those not attending school. The levels of public investment are necessarily much lower given other national investment priorities. Similarly, but at a smaller scale, private provision of digital training is available in many urban areas of the global South. The level of public engagement with digital developments may be, however, lower in the global South which creates an additional challenge known in Francophone areas as "sensibili-sation" – promoting the potential usefulness and value of knowledge and skills about digital developments in order to build interest and demand for training, education and investment.

The level of public engagement, along with connectivity issues, creates a bottleneck in the demand for investment in digital enterprises. In some cases, the global South has participated in digital development through the creation of facilities or city zones with advanced communication capabilities able to engage with international markets for services such as content moderation, call centres, voice transcription or software mainte-nance (Heeks, 2018). Although such enclaves may mitigate local short-comings in connectivity and contribute to skills by offering on-the-job training, they also exacerbate local inequalities and exclusion. Building local digital enterprises is a bigger challenge since it requires not only technological capabilities, but customers able and willing to change their business processes and methods (Graham, 2019). While it can be said that investment in such enterprises is lacking, so too is the demand for their potential services, creating a persistent chicken-and-egg problem.

7.3.1.3 "Accommodating" to local context and regulation

The phrase "accommodating to local context" spans a broad set of issues. The platform developments discussed in previous chapters emerged in a context where there is a defence of freedom of expression, a distaste for the state and a generally cosmopolitan view of the world in which differences in culture or politics are expected, but trivialised. For example, Facebook ardently maintains a commitment to free expression even when this involves promoting false and defamatory claims and refuses to exert (or erratically exerts) moderation on advertisements associated with government office seekers. This is extremely problematic in a global South context for a myriad of reasons. These include differences in cultural, social and political norms as well as the gendered experience of the digital environment which can yield harms and violence, especially for women, as can racial and ethnic tensions.[5] It is not only "freedom of expression" but also norms and values with regard to societal institutions that generate conflict between the "tech culture" of the "libertarian" West Coast of the US and other places in the world. Measures to address these kinds of conflicts have involved efforts to exert control over platform operations. These include choices of technology allowing states to shut down the internet within countries often in response to political protest,[6] and other restrictions on platform operations, all of which are contested.[7]

National regulation offers opportunities to influence platform company operations in the global South, although this requires resistance to norms and rules established in the global North. This resistance is not yet a prominent source of mitigating measures, partly because of acceptance of platform companies as embodying the future, or "modernisation", which is seen as desirable by some actors in the global South. In the case of the gig economy in Sri Lanka, for example, with some exceptions, microwork/online-freelancing is a largely unregulated second source of income. It is done in addition to a full-time job, but, at the same time, it is a growth opportunity for the country (Galpaya et al., 2018). Sympathies with modernisation mean that those already participating in platform services may be relatively indifferent to, or even actively work against norms transferred from the global North. This may be a result of an absence of capabilities to create local investment strategies, complicity with powerful local elites and corrupt governments, or the promotion of catch-up strategies without listening to what is being said locally to technology developers, aid agencies and companies (Manyozo, 2017).

"Accommodating to local context" involves a broad and often contradictory process of adaptation to language, symbols, conventions and implicit understandings that are part of local identities. Even if local investment is modest, there are possibilities for social networking and e-commerce that would better observe local cultural norms and rules. More controversially, an emphasis on local development may involve compromising the protection of rights well established elsewhere. For example, a Zimbabwean strategic cooperation agreement with a Chinese start-up provides for the export of a database of Zimbabwean citizens' faces to China in return for the country's access to computer vision technology.[8]

The rationales for closing digital divides or gaps are complicated by assumptions about whether improved connectivity, digital skills and policy adjustments are likely to lead to inclusion that is broadly beneficial, or whether it inevitably means new forms of exploitation. Alternatively, inclusion, even under exploitative conditions, may provide a "platform" for emancipation through resistance strategies that employ some of the tools of digital platforms.

7.3.2 Leapfrogging strategies

Leapfrogging traditionally means advancing along a linear and technologically convergent development pathway but skipping some of the intermediate steps. These intermediate steps may be unnecessary because of advances in knowledge and practice elsewhere in the world that can be observed and imitated or copied (Ernst et al., 2014; Soete, 1985). In other words, leapfrogging involves "bypassing some of the processes of accumulation of human capabilities and fixed investment in order to narrow gaps in productivity and output that separate industrialized and developing countries" (Steinmueller, 2001: 194). Questions of technology access and capabilities to make the leap are central to the potential for leapfrogging. In addition, leapfrogging requires capabilities to absorb and adapt technologies from elsewhere, a learning capacity that may be difficult to execute in practice (Steinmueller, 2000).

An alternative view of leapfrogging questions the destination of the leap. Innovation scholars who share some of the perspectives from each of the economic lenses (neoclassical, institutional and critical political economy) have observed that it is not only the rate of technological progress (along established trajectories of improvement), but also the *direction* of change that matters (Nelson, 1962; Stirling, 2007). At first glance, the view that one might leap in a different direction seems a hazardous undertaking; a leap into the unknown. It cannot be known whether this alternative direction is a dead end or, worse, would contribute to a decline in the conditions of the people in a society. What can be said is that the leap might be shaped by developing a different vision of the purposes for digitisation and datafication.

For example, the institutions (norms, rules and standards) governing the internet might be reformed to make companies posting user content legally liable for statements these users make. This would seriously diminish the prospects for social media companies and could lead to these companies abandoning a country implementing such a rule. Another possibility would be requiring platform companies to compensate users for participating on their platforms. This would create a cascade of user entrants motivated by the compensation more than by interest in platform participation. These examples would involve changes in prevailing platform practice, but they would transform the nature of datafication fundamentally, dramatically changing the power relationship between states, users and platform companies.

Less dramatic possibilities for leapfrogging to a different destination can be observed in the development of access provisions, such as community WiFi as an alternative means of connecting businesses and households. Community networks typically involve collective, or even family, ownership, management, open designs and open participation. They may aim to address issues of security and privacy or the promotion of content in local languages.[9] However, community networks can face major regulatory and financial burdens, as illustrated by a survey of such networks in Africa, finding that of 372 initiatives in 12 countries, only 25 were partially active.[10]

While these community networks do provide connectivity, they often are deemed by telecommunication and mobile operators to be uneconomic. Reports aimed at addressing the lack of connectivity focus principally on private initiatives such as Alphabet's Loon balloons or Facebook's Free Basics and it is difficult to shift attention to community-led initiatives. These community initiatives may seem a modest variant of the existing technological trajectory. However, combined with other initiatives such as limiting the scale of service suppliers and bundling in local services, there is potential for a positive effect. Even small changes in directionality can accumulate if subsequent organisational and institutional norms and rules reinforce a change in direction.

Countries in the global South, additionally, find that the race to digitalise is reinforcing their linkage to international value chains for better and for worse. International connections bring new employment opportunities but may displace local market development. Deliberating upon different directions might lead to different choices about technology, changes in the boundary between public and private, and in the regulation of the national operations of the foreign-owned (and domestic) platform companies. The uncertainties and challenges of leapfrogging would remain, but the direction of travel might better serve the interests of individuals and groups within the global South.

7.4   China's digital vision

In this section we consider digital platform developments in China where a mix of accommodation strategies and some of the features of leapfrogging measures are addressing their digital future through a combination of visions shared with, and diverging from, those of the global North. China has implemented various mitigating measures. It has achieved leapfrogging development in several areas successfully along conventional lines and it is promoting a different vision for the future of digitisation.[11] Its distinctive pattern of digital technology and institutional innovation is often seen as authoritarian, but this polemically oversimplifies the complexity of the economic, political and socio-cultural system that has positioned China's large digital platforms second only to the US.

China has attained this position by relying on a mix of enterprises, some domestic and some foreign, operating within a set of complex, often government-inspired, governance arrangements (Fuller, 2016). Its development of technologies and services benefits from state policy and regulation as well as local market knowledge which, in turn, benefits from a combination of protectionist policies that keep many foreign owned companies at a distance while, simultaneously, encouraging inward investment. The result is its distinctive form of "state-guided" capitalism (Yu, 2017).

The number of Chinese internet users surpassed the US in 2009 and in 2019 there were some 854 million users.[12] The Chinese internet governance system relies on an extremely complex and ever-changing mesh of institutions (Yang and Mueller, 2019). The country is home to very large and high-profile platform companies which benefit from substantial investment in the connectivity infrastructure, state support and access to international financial capital (Jia and Winseck, 2018). Prominent platform companies include Baidu (search and AI), Alibaba (e-commerce, digital entertainment and the cloud), Tencent (internet-related services including WeChat, entertainment content and technologies) and JD (e-commerce). Each of these companies is engaging successfully in technological developments in its principal area of operation by imitating and improving on pre-existing technologies and platform models.

The policy and regulatory environment adopted by China has favoured the growth of domestic companies and this is an instance of a mitigating measure in the framework discussed earlier in this chapter. Displaced multinationals claim that some of these measures are inconsistent with international agreements governing trade and investment. Claimed discrimination against foreign platforms includes a ban on Facebook because of its unwillingness to conform to regulatory requirements regarding state access to data. Google left the search market so as not to have to comply with these requirements. Apple transferred its iCloud data about Chinese users to a Chinese data centre to meet China's regulatory requirements. Although the Chinese platforms are growing rapidly – often through mergers and acquisitions – most have not adopted the international expansion strategies of Amazon, Facebook or Google. Without a change in strategy, it seems likely that most of these companies will not present a direct competitive threat to the dominant Western platforms for some time, despite the fact that there is increasing interdependence in terms of the trade in data flows (Mueller and Grindal, 2019).

Some Chinese platform companies are aiming to establish themselves as competitors in Western markets. TikTok, the short-form mobile video hosting platform, is owned by ByteDance, a large Chinese technology company. TikTok operates in some 150 markets and in at least 75 languages, and its operating norms and rules are attracting attention in the Western countries in which it operates.[13] It faces challenges in localising its content moderation practices to avoid charges of censorship in these markets and to ensure that its data collection and privacy protection practices comply with local expectations. The Alibaba Group's operations in e-commerce (business to business and retail) are traded on both the New York and Hong Kong stock exchanges. It is expanding into emerging economies as well as into Europe and the US. These operations confront the need to meet service quality standards and to avoid charges of serving as a hosting platform for counterfeit goods or concerns about their data collection practices.

An approach to leapfrogging is evident in China's science and technology strategies which include a mix of innovation and imitation. China aims to become a "science and technology superpower" across many areas (Ding, 2018). In the AI field, the aim is to achieve a world-leading level of capability in all aspects of AI – theory, technology and applications – by 2030.[14] The strategy is a state initiative, but dozens of agencies, private companies, academic institutions and subnational governments are involved. In one sense, a "catch-up" strategy, China's AI and platform-related strategies represent a distinctive innovation pathway, a leap in a different direction. There are, however, some darker elements. China's investment in AI and machine learning and its capabilities for launching intelligent agent "bots" and dis- or mis-information campaigns mean that the country is regarded in Western countries as a source of destabilising online content.

There are similarities to the West (the use of national champions, promotion of trade deals and investment in supercomputing). But there are also differences, especially regarding access to data, the strength of state support and a focused and proactive investment in a skilled labour force.[15] China is also leading discussions in international forums such as the International Telecommunication Union, proposing changes in the internet governance regime to secure national sovereignty. These are resisted by Western countries fearing state interventions in the global (and ideally open) internet. For example, China's proposal for "national cyber-sovereignty" is interpreted as requiring Chinese control over data repositories and the application of a strict liability regime for the digital platforms (Shen, 2016). Characterised as having the potential to create a "splinternet",[16] a relatively closed space, these measures confound several of the norms and rules that have previously governed the internet.

There is considerable attention in the West to Chinese policies aimed at controlling online communication. For example, a regulation bans user-generated audio and video content about current events. State control of the internet using the "Great Firewall" is intended to enforce censorship by using automated systems and humans to filter or block access to restricted foreign websites. The "Golden Shield" regulates information on domestic websites. Social media platforms are expected to require their users to be identifiable by state authorities. The information control system is not transparent and yields unpredictable results about what online content will be censored. The claim in the West is that the overall effect is to chill expression, although the magnitude of such chilling effects as compared to other non-Western countries – and certain trends towards censorship in the West – is hard to assess for observers outside China. The absence of an active public discourse makes it difficult to judge citizens' views on the legitimacy of commercial datafication and surveillance practices. Despite these concerns, the Chinese government has introduced standards for the protection of personal information which have been described as more restrictive than the European Union's GDPR.[17]

Chinese government efforts to control the internet are expected to play a key role in maintaining social stability and to integrate a wide range of public services, one of which is China's social credit systems. The social credit system aims to monitor and evaluate the activities of Chinese citizens, ranking their "trustworthiness" and producing a score(s) that can affect citizens' eligibility for public services, job offers or school access. Still in the early phase, The Ministry of Public Security is said to be building the world's largest facial recognition database and is experimenting with surveillance techniques. These measures may suggest a trajectory for future platform development elsewhere. Ambitions of the US-owned platforms for the advance of commercial datafication are following a similar path, albeit subject to more explicit patterns of opposition and resistance.

The way in which the digital platforms and the internet operate in China cannot be separated from the wider political and economic context or from the cultural conditions. In traditional journalism and in the wider media sphere, there is simultaneously consensus and contestation over how the power of the state operates in Chinese society. Ambivalence characterises the Chinese state's approach to the connectivity infrastructure and digital content. Meng (2018) argues that neither Chinese exceptionalism nor a liberal critique of China's policies on their own fully explain the way China's companies are being integrated into global capitalism. In this sense, the development of the internet, investment in AI and the support for digital platforms are battlefields for ongoing ideological contestation in China as they are elsewhere.

Examples of the complexity in reconciling the self-publishing affordances of the internet with control of dissent are visible in responses to online activism. In some instances, platform services are used by the Chinese state to provide early warning of potential social or political unrest; in others, controls are used to actively suppress debate among citizens. Yang (2017) argues that the state seeks to "demobilise" emotions associated with online protest, less through coercive methods than through proactive efforts to produce "positive energy", which supports the state's agendas.

The mix of measures adopted in China makes it a unique case in the development of digitisation and datafication. The combination of market size, apparent political unity and entrepreneurial initiative has enabled the country to set a path that departs in some ways from patterns of digital platform development in the West. China's success in creating its own platform companies and some of its efforts to control content are models that may be attractive in global South countries, especially when accompanied by investment in connectivity infrastructures. These developments may begin to influence future changes in the global North as well.

7.5   Conclusion

Asymmetric power matters. History shapes the way digital inclusion and exclusion operate through cultural, social, political and economic processes and inequalities. Economic outcomes related to digital technology are experienced differently depending upon people's life circumstances, institutional contexts and the configuration of power relations. The conditions under which it might be possible for people in the global South (and elsewhere) to acquire capabilities to shape their own technological environments in ways that are beneficial, instead of unsustainable and destructive, do not arise simply from the "disruptive" force of digital technologies. It is not the newness of digital platforms that poses the issues discussed in this chapter, but a persistent reluctance to take seriously the fact that, like other technologies, digital platforms and other Fourth Industrial Revolution developments are means of "producing instruments of control and influence over other individuals, firms and nations".[18] This was acknowledged in the 1970s, but this is still absent from much of the contemporary discussion.

In leapfrogging, falling-behind and catching-up scenarios, even if they require a struggle, there are large numbers of alternative choices that can be taken in the global South and North. Some of these are illustrated in our examination of China's experience with digitalisation and platforms. What roles technologies play and how leadership is defined ought to vary in response to location, conditions and priorities, with the caveat that there may be broad, if not universal, agreement about certain principles and rights. Alternatives cannot be imagined and implemented effectively, however, without a good understanding of why dominant platforms have emerged, which interests are promoting their further commercial development, and what (even if occasional) opportunities there are to start developing them differently.

In the next and final chapter, we summarise the contributions of economic analysis to the opportunities and problems presented by the digital platforms and consider the road ahead.

Notes

1. See WEF (2017, 2019).
2. OECD (2015).
3. Mapping using UNESCO's Internet Universality Indicators seeks to provide contextual views of performance, while the Digital Rights Index measures corporate accountability, see Souter and Van der Spuy (2019) and Ranking Digital Rights (2019).
4. See Castells and Himanen (2014), Hesmondhalgh (2017) and Kleine (2013) for elaborations of alternative views.
5. See Banaji et al. (2019).
6. See Taye and Cheng (2019) for internet shutdowns by country 2018.
7. See Alemany and Gurumurthy (2019).
8. See Jie (2018).
9. See Belli (2017).
10. See Moreno-Rey and Graaf (2017) and Kretschmer et al. (2019).
11. See Roberts et al. (2019).
12. Some 541 million people or 37.2 per cent of the population were non-internet users, and some 14.2 per cent of these non-users were not online due to their age (too young or too old), see CNNIC (2019: 17).
13. In 2019, ByteDance was fined USD 5.7 million by the US Federal Trade Commission for TikTok's collection of personal data of children under the age of 13 without consent. This is being investigated by the UK Information Commissioner's Office with a decision pending at the time of writing.
14. See China Innovation Funding (2017).

15. See Ding (2018).
16. See Morozov (2010).
17. See Bolsover (2017).
18. See Dag Hammarskjöld Foundation (1975: 93) with echoes in the work of Bell and Pavitt (1997).

8. Conclusion

8.1 Introduction

Digital platforms are predominant means of offering social media, e-commerce, publishing and audiovisual online services as well as information services such as internet searching. Our principal focus in this book is on the largest platforms. This is because the services they offer are reaching into and affecting the lives of the greatest number of people. We focused on the behaviours and practices of the companies operating under private sector incentives because these incentives inflect and steer the creation and delivery of services in ways that create conflicts with public values. These conflicts are not attributable solely to technology or to the predominance of the advertiser-supported business model that thrives on commercial datafication. Platform strategies are inseparable from practices that reinterpret and reshape societal norms and rules – practices which are reflections of power asymmetries in society.

Public concerns about platform "bad behaviours" are fuelling a climate of "regulation by outrage" with potentials for policy and regulatory outcomes that could either benefit or harm consumers or citizens. Better outcomes and public understanding can benefit from a deeper economic analysis of how and why platforms evolved to their present state and what future they offer. In the preceding chapters, we discussed three traditions in the economic analysis of platforms – neoclassical and institutional economics and critical political economy. Each one contributes to a more complete understanding of the origins and consequences of the current configuration of platform strategies and practices.

So, what does economic analysis tell us about the benefits as well as the harms that arise because of the ascent of platforms? How might the benefits and harms be influenced by the choices of platform operators, policy makers or other actors?

8.2   Revisiting the digital platform problem

Digital platforms are a major or radical innovation because of their application to many different human activities. Like other major innovations, their developers had to discover a means to translate their potential into the reality of actual products and services. We have noted that major innovations are shaped by the context in which they first appear but that their widespread take-up involves changing this context through further complementary innovations and by restructuring or reinterpreting the norms, rules and standards of society. The internet environment from which these platforms emerged was a data communication infrastructure that was created to facilitate scientific and technological research, largely in the government and university sector. This context involved communication among people employed by these institutions. In this context, a relatively high degree of decorum and civility was a norm reinforced by codes of practice and potential sanctions against transgressive behaviour. This environment was changed fundamentally by a set of policy decisions allowing commercial use of the internet. This decision was taken with very little knowledge of what might ensue, other than that it was expected to open business opportunities in an emerging digital economy.

The digital service pioneers confronted a fundamental economic problem. How could they make money from information on a system designed to freely exchange information? Economics, in all its manifestations, wrestles with the problem of scarcity. There certainly were some elements of scarcity – most businesses and almost all consumers did not have the connectivity available in universities and research labs. Money was made in extending the internet to homes and previously unconnected workplaces. This made it possible to do many useful things such as writing e-mails and viewing the pioneering information services that made some money by including advertising. The World Wide Web, another non-commercial innovation, improved the range of information by making a standard for finding it. It was at this juncture that the key innovations emerged that made digital platforms possible. Google and other companies realised that they could learn about their users and use this information to sell higher-value advertising because it could be targeted at user interests. From this key insight, a cascade of complementary innovations followed to monetise user data and observational data about users – the process now referred to as datafication.

How did these companies know who their users were? The earlier scientific and technical research institutional context was one in which the norm was openness and there was an absence of concealment of identity with a relative indifference to privacy. These norms were embedded in the software that made the internet work. Strictly speaking, this was not necessary. An internet which preserved anonymity and privacy would have been possible, but this did not happen largely because of its origins in the research community. The internet's technical design and protocols are a consequence of human choice, not a technical necessity.

The platforms are providing access to digital content that is desired by some users and not by others. Some of this content is illegal and some is deemed harmful. The dominant platforms employ datafication in increasingly sophisticated ways, using AI-enabled predictive engines to identify but also to shape user viewing behaviour. The platform model is highly disruptive, squeezing incumbent companies ranging from "brick and mortar" retailers to newspaper and magazine publishers. For many economists, this disruption is an expected competitive consequence of innovation; winners and losers from competition are determined by customer choices. The remarkable growth, and now enormous size, of some platform companies have allowed them to diversify into auxiliary services such as the server farm-enabled cloud, AI-as-a-service and data analytics services. AI-enabled prediction machines in e-commerce are credited with efficiency gains and these machines serve as control systems for managing public health and environmental degradation. Professional news organisations used to be the main filter through which the public received information, whereas more diverse voices can now be heard. For those with a sufficiently high level of critical literacy, these changes are positive and offer better opportunities to participate in democratic society.

Yet these platforms are also associated with an unusually extensive collection of negative externalities or social, political and economic harms. In seeking to capture user attention, they nudge individuals to interact online and offline in ways that can be harmful to themselves and others. The platforms' operations are associated with multiple societal problems.

There is growing discomfort with the fact that by virtue of their size, strategies and practices, the largest platforms are eroding the capacities of the state or individual users to exercise control over their datafication operations. Inconsistencies between the platforms' stated purposes and behaviours and public values that are of importance in democracies are leading the platforms to enact new forms of self-regulation and to external policy and regulatory initiatives. This trend is likely to continue with the consideration of combinations of measures involving state/corporate co-regulation, state-only regulation and independent regulatory institutions. These external measures will be opposed by the platforms and treated as interference and constraints on their freedom to reap a return on their investments.

The outcomes of new policy and regulatory measures, including whether they overcome resistance, will depend partly on the specific norms and rules that are introduced and how they are enforced. All solutions, including independent institutions with diverse actor representation embracing civil society, will have to address ambiguities around what is defined and operationalised as harmful content, practice and behaviour – that is, how they understand the always contested public interest. If enacted, the outcomes will depend on the adequacy of knowledge and the political will to enforce new norms, rules and standards. Many, including the platform owners, will argue that self-regulation and the potential for competition, domestically or internationally, are better solutions to remedying perceived shortcomings than external interference. Some external reforms may lead to positive change, others may lead to worsening conditions. Notwithstanding this uncertainty, the three traditions in economic analysis highlight how arguments for different approaches are likely to be presented.

- Improving competition is a principal policy prescription arising from a neoclassical analysis of platforms and often from both the newer and older institutional traditions. The case for market intervention through merger and acquisition control or antitrust proceedings typically, however, is weak. The complexity and rapid evolution of platforms and their ecologies make demonstration of actionable offences difficult. Analysis offers many alternative interpretations of what might follow from market intervention. The business model of the multisided platform diverges from the common assumption that markets are composed of atomistic sellers. The nature of a "market" in which an intermediary can set the terms of exchange and gain knowledge about buyers and sellers that allows it to shape their decisions challenges foundational assumptions used in neoclassical analysis. Public values are only obliquely addressed although there are signs that privacy considerations may be brought within this framework. Keeping to the basics, however, a common argument will be that platforms, whatever their size, are disciplined by the prospect of a mass exodus of customers and, thus, must treat them fairly. Arguments that the largest platforms have achieved an unreproducible position, a "natural monopoly", or are an essential facility or infrastructure, will be countered by claims that innovation continues to be dynamic in creating new competitive alternatives and that their very size is evidence of their innovative success. The numerous corollaries to this argument provide a means of justifying mergers and acquisition and unusually high profits compared to other businesses, and a basis for rebuffing other complaints.

- A "new" institutional economic analysis is open to investigations of the rules by which platforms achieve their dominance. However, like neoclassical economics, these rules are taken to be determined externally. The purpose of an economic analysis here is to provide an understanding of the possible consequences of alternative rules, not to offer normative prescriptions for how those rules should be changed. Both this and the neoclassical approach focus on how restrictions might hinder platform expansion and provide rivals with advantages that might jeopardise innovation. Policy responses are likely to take a promotional approach to the construction and operation of platforms without fundamentally challenging the commercial data-fication model. The consideration of institutional norms and rules in the new institutional economics tradition may, however, make it more open to findings of anticompetitive practices. It may, therefore, provide a basis for stronger advocacy of regulatory constraints on the platforms' practices or even structural remedies such as the break-up of the dominant companies.

- In other (older) institutional economic traditions, a focus on platform operations and their misalignment with public values provides a stronger and often normative path to regulatory reform. Insights into how asymmetrical power relations are complicit in yielding problematic outcomes can support proposals for use of the tools of governance by the state, co-regulation or independent regulation to change the incentives influencing platform behaviour to make the platforms accountable to the public. Regulatory responses have the potential to encourage a shift in the boundary between public and private platform provision towards the public and to suggest alternative operational models. Challenges to the basic platform business model are likely to suggest that alternative business models should operate adjacent to existing models. These might be those we have discussed – subscription, public service, club and voluntary contribution models – or they might be other models that will be created through future imaginative innovations.

• In contrast, a critical political economy analysis of digital platforms focuses from the outset on the way they unjustly expropriate value from users. It is likely to call for more far-reaching change in the norms and rules governing platform operations. From this perspective, the inadvertent contribution of economic value to the platform, generated by users' contributions and observation of them, is an exploitative practice of contemporary capitalism. The central insight of this approach is the nature of appropriation of individual and collective value creation by owners with meagre or no returns to users. Redressing this appropriation (or expropriation) requires altering the boundary between what may be owned and utilised for commercial purpose and what should be considered collective or public property. Doing this may involve taxation or public ownership, aiming to return ownership and control to platform users.

Both the institutional economic and critical political economy traditions question the desirability of a future in which everyone simply adjusts to the radical innovations that have given rise to commercial platforms. They call into question the claimed inevitability of the commercial datafication business model. They aim to evaluate alternatives that might better uphold public values – alternatives that might give people "freedom from" the commercial logic of datafication if they wish and "freedom to" create alternatives (Milan, 2019). All three of these economic traditions acknowledge that there may be unintended (negative as well as positive) outcomes if new norms and rules can be imposed on the platforms. They differ in the extent to which they advocate a normative position based on their analysis.

8.3   The road ahead

Policy and regulatory measures are being debated because of growing concern about the largest digital platforms' power to operate without public accountability. Negative consequences are seen by policy makers as outweighing the positive gains for individuals and for democracy. Interests in enhancing security to guard against some of the platforms' and users' activities that are illegal or meet with disapproval struggle against the interests of those defending the right to be free from surveillance or to uphold other public values. The impetus for action includes panics or outrage about infringements of human rights, the decline of incumbent companies such as commercial news publishers and public service media. Policy and regulatory reforms are political processes. They respond in unpredictable ways to a mix of outrage and evidence. Policy and regulatory intervention may involve a shift in the locus of governance power or in the boundary between the public and private supply of platform services. Shifts towards public supply are more likely in an environment with an increasingly toxic digital landscape with ever-diminishing privacy and the ascendance of AI-enabled applications, beyond the reach of individual and, often, collective control.

If, in the coming years, we find that in healthcare, education, finance, transport or the conduct of wars, decisions increasingly are made based on opaque calculations, then no one will have decided, other than to have taken a decision to cede control to the platforms' systems. Will this ceding of human volition matter if everyone is better off than they are today? What if everyone, or some groups, are much worse off than they are today? The choices made in governing the digital platforms are also choices about where the moral limit of the commercial market should be set in Western democracies and elsewhere (and about the limits and restraints on the liberal state).[1] In our view, these choices need to err towards creating sustainable opportunities for imagining and experimenting with alternatives.[2]

Neoclassical economic analysis emphasises that radical innovations like digitalisation or datafication require new or adjusted self-regulation practices. Enabling legislation allows private enterprises to take advantage of new potentials. The other two traditions in economic analysis emphasise that neither digitalisation, nor disruptions to public values in the form of commercial datafication, are inevitable. There are choices to be made about the design and deployment of these technical systems. Choices are being enacted at multiple scales – local, national and international. These include choices embedded in codes of practice, in the operations of algorithms and about acceptable market structures. They centre around datafication processes. In our view these choices should be overseen by participatory and democratic institutions, independent of platforms and governments.

Oversight institutions of policy and regulation charged with responsibilities for upholding public values, assuring transparent and
non-discriminatory business practices, and protecting the rights of citizens and consumers, exist in other sectors. Invariably, these institutions require the production of data on the operation of enterprises.
This requires the development of professional staff with the technical
knowledge to critically assess the operations of their industry sector.
These institutions are always at risk of capture or undue influence by the
enterprises they are charged with regulating but can be held to account
by democratic processes. Well-functioning regulatory institutions are
difficult to achieve, but not impossible to imagine.

Some will conclude that the largest digital platforms represent an unacceptable level of control in specific markets or in relation to citizens and
will propose structural remedies such as breaking up or nationalising
platform companies. In the first instance, it must be observed that
such structural remedies involve jurisdiction. Despite their multinational
operations, the largest platform companies are American and Chinese.
Some structural remedies would have to be implemented by the political
and judicial systems of those countries. Secondly, structural remedies
have had mixed results, including many unintended consequences. For
example, the break-up of AT&T in the 1980s in the US produced benefits
for telecommunication customers, it stimulated innovation in data processing domains, but it also destroyed Bell Labs, an outstanding corporate
research institution. The level and variety of research being conducted
by the largest platform companies suggests similar risks and benefits in
undertaking structural solutions. These measures may force changes in
the internal standards and practices of the platforms or the unbundling of
divisions of the largest platforms, but there is also a risk that the companies will opt for compliance rather than any fundamental change in their
basic datafication business model.

In our view none of these measures is likely to secure a very substantial
shift towards public interest platform operation or a marked elevation of
care for public values. This would require an effort to secure the financial
sustainability of alternative business models that do not operate with the
principal goal of profiting from datafication or structural remedies that
make one or more platforms public enterprises or organisations. A move
away from the advertiser-supported business model, either wholly or
far more substantially than in the present, means rethinking how digital
technologies and systems should be financed and deployed. Codes of

good (ethical) behaviour or the break-up of dominant platforms are not likely to alter the progress of commercial datafication and the further infiltration of algorithms into decision making in society. For a significant change to happen, people must be able to exit commercial datafication, but they can only do so if they have somewhere to exit to. In addition, any public initiatives using processes of datafication "for good" will also need to be subject to internal or external oversight to ensure that public values are being upheld in practice.

All these choices invoke public values – not only economic value – considerations. In one scenario, it may be concluded that the advance of AI-enabled datafication practices is valued in Western democracies, but that public ownership and investment in non-commercial business models are essential, e.g. publicly funded search engines, collectively owned data repositories that do not sell data or use it to generate advertising revenues, and well-funded public service media. There would still be a requirement for effective public interest governance of how and by whom user-contributed and user interaction-generated data are managed. In another scenario, it may be concluded that it is not feasible to operate AI-enabled applications with the degree of accountability that is required to uphold public values. Certain applications, whether in commerce or public management, may be deemed unacceptable and halted by law. Regardless of whether this scenario prevails in Western countries, other countries are likely to continue to pursue more intensive uses of surveillance and the application of technologies of citizen control.

The most realistic scenario in the medium term is: (1) implementation of conventional tools of regulation through a mix of self-regulation and new regulatory oversight institutions; (2) a ban on the use of datafication in particularly sensitive areas such as politics or content directed at children; and (3) investment in alternative business models substantial enough to enable the provision of platform services respecting public values at scale. The contemporary information crisis may be less prominent than other global crises such as climate change or poverty, but it deserves attention and major investment in a platform infrastructure consistent with public values in democracies and with public service media to enable informed discussion.

The assumption must be abandoned that individual children, youths and adults operate with the agency required to keep themselves safe and informed to standards expected in a democracy. Digital literacies are essential for navigating whatever digitalised world emerges, but they cannot compensate for the absence of collectively inspired changes in the norms, rules and standards for digital platform operation. Collectively inspired changes are also essential to support workers and their conditions of employment. In all these areas, there is no likelihood that a consensus about the necessary changes will emerge easily.

8.4 Conclusion

Economic analysis reveals that the prominence of the commercial platform mode of economic organisation is the result of economic, political and social choices that established the norms, rules and standards enabling the rise of dominant platforms. These platforms are delivering very substantial benefits in informing, educating and entertaining vast numbers of people; and many claim they are supporting human flourishing. However, economic analysis also yields evidence to implicate the commercial platform mode of economic organisation and its technologies of digitalisation and datafication in some of the negative societal developments in our time.

Judgements about the balance between benefit and hazard are not ours alone to make, but we have made our preference for reform clear. If digital technologies and platforms are to provide a foundation or infrastructure embodying human rights and yield fair and equitable outcomes, the norms, rules and standards governing them need to change. The algorithms being used to guide or supplant human decision making and the prominence of commercial datafication operations are being encouraged by the notion that technological innovation shapes the destiny of human beings. We join with Benkler et al. (2018: 381) in asserting that "technology is not destiny".

Notes

1. As Kant put it, "in the kingdom of ends, everything has either a price, or a dignity. What has a price can be replaced with something else, as its equivalent, whereas what is elevated above any price, and hence allows of no equivalent, has a dignity" (Kant, 1785/2012: 46).
2. See Mansell (2012, 2017) and Mosco (2019) for discussions of conditions needed for imaginaries to flourish.

References

All URLs last accessed 3 January 2020.

Acemoglu, D., Makhdoumi, A., Malekian, A. and Ozdaglar, A. (2019) *Too much data: Prices and inefficiencies in data markets.* NBER Working Paper 26296. Cambridge, MA. At https://www.nber.org/papers/w26296

Akerlof, G. (1970) 'The market for lemons, quality uncertainty and the market mechanism'. *Quarterly Journal of Economics, 84*(3): 488–500.

Al-Ani, A. and Stumpp, S. (2016) 'Rebalancing interests and power structures on crowdworking platforms'. *Internet Policy Review, 5*(2): 1–20.

Alemany, C. and Gurumurthy, A. (2019) 'Governance of data and artificial intelligence'. In B. Adams, C. Alemany, R. Bissio, C. Y. Ling, K. Donald, J. Martens and S. Prato (eds). *Spotlight on sustainable development 2019: Reshaping governance for sustainability – Global civil society report on the 2030 agenda and the SDGs* (pp. 86–94): Reflection Group on the 2030 Agenda for Sustainable Development. At https://www.2030spotlight.org/en/book/1883/chapter/reshaping-governance-sustainability

Algorithm Watch. (2019) *AI ethics guidelines global inventory.* AlgorithmWatch gGmbH. Berlin. At https://algorithmwatch.org/en/project/ai-ethics-guidelines-global-inventory/

Alphabet. (2018) *Alphabet annual report.* Alphabet Form 10-K, submitted to US Securities and Exchange Commission. Washington, DC. At https://abc.xyz/investor/static/pdf/2018_alphabet_annual_report.pdf?cache=61d18cb

Amazon. (2018) *Amazon annual report.* Amazon Form 10-K, submitted to US Securities and Exchange Commission. Washington, DC. At https://ir.aboutamazon.com/static-files/0f9e36b1-7e1e-4b52-be17-145dc9d8b5ec

Ananny, M. (2016) 'Toward an ethics of algorithms: Convening, observation, probability, and timeliness'. *Science, Technology & Human Values, 41*(1): 93–117.

Ananny, M. and Crawford, K. (2018) 'Seeing without knowing: Limitations of the transparency ideal and its application to algorithmic accountability'. *New Media & Society, 20*(3): 973–989.

Anderson, C. (2006) *The long tail: Why the future of business is selling less of more.* New York: Hyperion.

Arrow, K. J. (1984) *The economics of information: Collected papers of Kenneth Arrow, Volume 4.* Oxford: Blackwell.

Arthur, W. B. (2015) 'All systems will be gamed: Exploitative behavior in economic and social systems'. In W. B. Arthur (ed). *Complexity and the economy* (pp. 103–118). Oxford: Oxford University Press.

Atkinson, A. B. (2008) *The changing distribution of earnings in OECD countries*. Oxford: Oxford University Press.

Atkinson, A. B. (2015) *Inequality: What can be done?* Cambridge, MA: Harvard University Press.

Autor, D., Dorn, D., Katz, L. F., Patterson, C. and Van Reenen, J. (2020) 'The fall of the labor share and the rise of superstar firms'. *Quarterly Journal of Economics*, 135(2): 645–709.

Balkin, J. M. (2016) 'Information fiduciaries and the First Amendment'. *UC Davis Law Review, 39*(4): 1185–1234.

Ballon, P. (2009) 'The platformisation of the European mobile industry'. *Communications & Strategies, 75*(3): 15–33.

Banaji, S., Bhat, R., with Agarwal, A., Passanha, N. and Pravin, M. S. (2019) *WhatsApp vigilantes: An exploration of citizen reception and circulation of WhatsApp misinformation linked to mob violence in India*. Department of Media and Communications, London School of Economics and Political Science. At http://www.lse.ac.uk/media-and-communications/assets/documents/research/projects/WhatsApp-Misinformation-Report.pdf

Barocas, S. (2014) *Data mining and the discourse on discrimination*. Proceedings of the Data Ethics Workshop, Conference on Knowledge Discovery and Data Mining (KDD). At https://pdfs.semanticscholar.org/abbb/235fcf3b163af d74e1967f7d3784252b44fa.pdf

Barwise, P. and Picard, R. G. (2014) *What if there were no BBC television? The net impact on UK viewers*. Reuters Institute for the Study of Jouralism. Oxford. At https://reutersinstitute.politics.ox.ac.uk/sites/default/files/2017-06/What%20if %20there%20were%20no%20BBC%20TV_0.pdf

Bauer, J. M. (2014) 'Platforms, systems competition, and innovation: Reassessing the foundations of communications policy'. *Telecommunications Policy, 38*(8–9): 662–673.

Bauer, J. M. (2018) 'The internet and income inequality: Socio-economic challenges in a hyperconnected society'. *Telecommunications Policy, 42*(4): 333–343.

Bauer, J. M. and Latzer, M. (eds). (2016) *Handbook on the economics of the internet*. Cheltenham, UK and Northampton, MA, USA: Edward Elgar Publishing.

Baumol, W. J. (2004) 'Red-Queen games: Arms races, rule of law and market economies'. *Journal of Evolutionary Economics, 14*(2): 237–247.

Bayamlioglu, E., Baraliuc, I., Janssens, L. and Hildebrandt, M. (eds). (2018) *Being profiled: Cogitas ergo sum 10 years of profiling the European citizen*. Amsterdam: Amsterdam University Press.

Beckett, C. and Mansell, R. (2008) 'Crossing boundaries: New media and networked journalism'. *Communication, Culture & Critique, 1*(1): 90–102.

Bell, E. and Owen, T. (2017) *The platform press: How Silicon Valley reengineered journalism*. Tow Center for Digital Journalism, Columbia Journalism School. At https://www.cjr.org/tow_center_reports/platform-press-how-silicon-valley -reengineered-journalism.php

Bell, M. and Pavitt, K. (1997) 'Technological accumulation and industrial growth contrasts between developed and developing countries'. In D. Archibugi and J. Michie (eds). *Technology, globalisation and economic performance* (pp. 83–137). Cambridge: Cambridge University Press.

Belli, L. (2017) *Community networks: The internet by the people, for the people: Official outcome of the UN IGF dynamic coalition on community connectivity*. Internet Governance Forum. Geneva. At http://bibliotecadigital.fgv.br/dspace/bitstream/handle/10438/19401/Community%20networks%20-%20the%20Internet%20by%20the%20people,%20for%20the%20people.pdf?sequence=1&isAllowed=y

Benkler, Y. (2006) *The wealth of networks: How social production transforms markets and freedom*. New Haven, CT: Yale University Press.

Benkler, Y. (2016) 'Peer production and cooperation'. In J. Bauer and M. Latzer (eds). *Handbook on the economics of the internet* (pp. 91–119). Cheltenham, UK and Northampton, MA, USA: Edward Elgar Publishing.

Benkler, Y. and Nissenbaum, H. (2006) 'Commons-based peer production and virtue'. *Journal of Political Philosophy, 14*(4): 394–419.

Benkler, Y., Faris, R. and Roberts, H. (2018) *Network propaganda: Manipulation, disinformation, and radicalization in American politics*. Oxford: Oxford University Press.

Bennett, C. J. and Raab, C. D. (2018) 'Revisiting the governance of privacy: Contemporary policy instruments in global perspective'. *Regulation & Governance*, first published 27 September: 1–18.

Berger, T. and Frey, C. B. (2016) *Digitalisation, deindustrialisation and the future of work*. OECD Social, Employment and Migration Working Papers, No. 193. Paris. At https://www.oecd-ilibrary.org/docserver/5jlr068802f7-en.pdf?expires=1578136825&id=id&accname=guest&checksum=A0128025898B7D4A8E04A815B45C5517

Black, R. E. (2012) *Porta Palazzo: The Anthropology of an Italian market*. Philadelphia, PA: University of Pennsylvania Press.

Bodó, B., Helberger, N., Eskens, S. and Möller, J. (2019) 'Interested in diversity: The role of user attitudes, algorithmic feedback loops, and policy in news personalization'. *Digital Journalism, 7*(2): 206–229.

Bolsover, G. (2017) *Computational propaganda in China: An alternative model of a widespread practice*. OII, Computational Propaganda Project, Working Paper, No. 2017.4. Oxford. At https://www.oii.ox.ac.uk/blog/computational-propaganda-in-china-an-alternative-model-of-a-widespread-practice/

Bork, R. H. and Sidak, J. G. (2012) 'What does the Chicago School teach about internet search and the antitrust treatment of Google?' *Journal of Competition Law & Economics, 8*(4): 663–700.

Bourgine, P. (2004) 'What is cognitive economics?' In P. Bourgine and J. P. Nadal (eds). *Cognitive economics* (pp. 1–12). Berlin: Springer.

Bowker, G. C. and Star, S. L. (1999) *Sorting things out: Classification and its consequences*. Cambridge, MA: The MIT Press.

Bozeman, B. (2002) 'Public-value failure: When efficient markets may not do'. *Public Administration Review, 62*(2): 145–161.

Bozeman, B. (2007) *Public values and public interest: Counterbalancing economic individualism*. Washington, DC: Georgetown University Press.

Branscomb, A. (1994) *Who owns information? From privacy to public access*. New York: Basic Books.

Brautigan, R. (1983) *The abortion: A historical romance 1966*. New York: Picador.

Brennen, J. S., Noward, P. N. and Nielsen, R. K. (2019) *An industry-led debate: How UK media cover artificial intelligence*. Reuters Institute, OII, Oxford Martin School, University of Oxford. At https://reutersinstitute.politics.ox.ac.uk/sites/default/files/2018-12/Brennen_UK_Media_Coverage_of_AI_FINAL.pdf

Bresnahan, T. and Greenstein, S. (2014) 'Mobile computing: The next platform rivalry'. *American Economic Review, 104*(5): 475–480.

Bresnahan, T. and Trajtenberg, M. (1995) 'General Purpose Technologies "engines of growth?"' *Journal of Econometrics, 65*(1): 83–108.

Brynjolfsson, E. and Kahin, B. (eds). (2002) *Understanding the digital economy: Data, tools, and research*. Cambridge, MA: The MIT Press.

Brynjolfsson, E. and McAfee, A. (2014) *The second machine age: Work, progress, and prosperity in a time of brilliant technologies*. New York: W W Norton.

Buchanan, J. M. (1965) 'An economic theory of clubs'. *Economica, 32*(1): 1–14.

Cammaerts, B. and Mansell, R. (2020) 'Digital platform policy and regulation: Toward a radical democratic turn'. *International Journal of Communication, 14*: 1–19.

Castells, M. (2009) *Communication power*. Oxford: Oxford University Press.

Castells, M. and Himanen, P. (eds). (2014) *Reconceptualizing development in the global information age*. Oxford: Oxford University Press.

China Innovation Funding. (2017) *State Council's plan for the development of new generation artificial intelligence*. State Council. Beijing. At http://chinainnovationfunding.eu/dt_testimonials/state-councils-plan-for-the-development-of-new-generation-artificial-intelligence/

Christensen, C. (1997) *The innovator's dilemma: When new technologies cause great firms to fail*. Cambridge, MA: Harvard Business School Press.

CIGI. (2019) *2019 CIGI-Ipsos global survey on internet security and trust*. Centre for International Governance Innovation and Ipsos. At https://www.cigionline.org/internet-survey-2019

Clark, D. D. (2018) *Designing an internet*. Cambridge, MA: The MIT Press.

CNNIC. (2019) *The 44th China statistical report on internet development*, trans. B. Meng. China Internet Network Information Center (CNNIC). At http://www.cac.gov.cn/2019-08/30/c_1124938750.htm

Cohen, J. E. (2016) 'The regulatory state in the Information Age'. *Theoretical Inquiries in Law, 17*(2): 369–414.

Couldry, N. (2019) 'Capabilities for what? Developing Sen's moral theory for communications research'. *Journal of Information Policy, 9*: 43–55.

Couldry, N. and Mejias, U. A. (2019) *The costs of connection: How data is colonizing human life and appropriating it for capitalism*. Standford, CA: Stanford University Press.

Council of Europe. (2018) *Recommendation on media pluralism and transparency of media ownership*. Council of Europe Recommendation CM/Rec(2018)1 of the Committee of Ministers to Member States. Strasbourg. At https://search.coe.int/cm/Pages/result_details.aspx?ObjectId=0900001680790e13

Crémer, J., de Montjoye, Y.-A. and Schweitzer, H. (2019) *Competition policy for the digital era: Final report*. Directorate-General for Competition. Brussels. At https://ec.europa.eu/competition/publications/reports/kd0419345enn.pdf

Culliford, E. (2019) 'Warren campaign challenges Facebook ad policy with "false" Zuckerberg ad', *Reuters*, 12 October. At https://www.reuters.com/article/us

-usa-election-facebook/warren-campaign-challenges-facebook-ad-policy-with
-false-zuckerberg-ad-idUSKBN1WR0NU

Dag Hammarskjöld Foundation. (1975) *The 1975 Dag Hammarskjöld report on development and international cooperation*. Dag Hammarskjöld Foundation for United Nations General Assembly, 7th Special Session. Motala. At https://www.daghammarskjold.se/publication/1975-dag-hammarskjold-report-development-international-cooperation/

David, P. A. and Steinmueller, W. E. (1994) 'Economics of compatibility standards and competition in telecommunication networks'. *Information Economics and Policy*, 6(3–4): 217–241.

Davies, H. C. and Eynon, R. (2018) 'Is digital upskilling the next generation our "pipeline to prosperity"?' *New Media & Society*, 20(11): 3961–3979.

DeNardis, L. (2014) *The global war for internet governance*. New Haven, CT: Yale University Press.

DeNardis, L. and Hackl, A. M. (2015) 'Internet governance by social media platforms'. *Telecommunications Policy*, 39(9): 761–770.

Ding, J. (2018) *Deciphering China's AI dream: The context, components, capabilities, and consequences of China's strategy to lead the world in AI*. Governance of AI Program, Future of Humanity Institute, University of Oxford. At https://www.fhi.ox.ac.uk/wp-content/uploads/Deciphering_Chinas_AI-Dream.pdf

doteveryone. (2018) *Regulating for responsible technology*. doteveryone. London. At https://doteveryone.org.uk/wp-content/uploads/2018/10/Doteveryone-Regulating-for-Responsible-Tech-Report.pdf

EC. (2000) *Directive on electronic commerce*. European Commission. Brussels. At https://eur-lex.europa.eu/LexUriServ/LexUriServ.do?uri=OJ:L:2000:178:0001:0016:EN:PDF

EC. (2002) *Directive on privacy and electronic communications*. European Commission, OJ L 201/37, 31 July. Brussels. At https://eur-lex.europa.eu/legal-content/EN/TXT/PDF/?uri=CELEX:32002L0058&from=EN

EC. (2009) *Application of state aid rules to public service broadcasting*. European Commission, OJ L 257/1, 27 October. Brussels. At https://eur-lex.europa.eu/legal-content/EN/TXT/PDF/?uri=CELEX:52009XC1027(01)&from=EN

EC. (2010) *Audiovisual media services directive*. European Commission, OJ L 95/1, 15 April. Brussels. At https://eur-lex.europa.eu/legal-content/EN/TXT/PDF/?uri=CELEX:32010L0013&from=EN

EC. (2016) *General data protection regulation*. European Commission, OJ L 119/1, 4 April. Brussels. At https://eur-lex.europa.eu/legal-content/EN/TXT/HTML/?uri=CELEX:32016R0679

EC. (2018a) *Audiovisual media services directive*. European Commission, OJ L 303/69, 28 November. Brussels. At https://eur-lex.europa.eu/legal-content/EN/TXT/PDF/?uri=CELEX:32018L1808&from=EN

EC. (2018b) *Recommendation on measures to effectively tackle illegal content online*. European Commission C(2018) 1177 final. Brussels. At https://ec.europa.eu/digital-single-market/en/news/commission-recommendation-measures-effectively-tackle-illegal-content-online

EC. (2019a) *Antitrust: Commission opens investigation into possible anti-competitive conduct of Amazon*. European Commission Press Release. Brussels. At https://ec.europa.eu/commission/presscorner/detail/en/ip_19_4291

EC. (2019b) *The changing nature of work and skills in the Digital Age.* European Commission Joint Research Centre. Brussels. At https://ec.europa.eu/jrc/en/publication/eur-scientific-and-technical-research-reports/changing-nature-work-and-skills-digital-age

Eide, M., Sjøvaag, H. and Larsen, L. O. (eds). (2016) *Digital challenges and professional reorientations: Lessons from Northern Europe.* Bristol: Intellect.

Ernst, D., Lee, H. and Kwak, J. (2014) 'Standards, innovation, and latecomer economic development: Conceptual issues and policy challenges'. *Telecommunications Policy, 38*(10): 853–862.

Eubanks, V. (2018) *Automating inequality: How high-tech tools profile, police, and punish the poor.* New York: St. Martin's Press.

European Parliament. (2019) *Amendments on directive on copyright in the digital single market.* European Parliament, P8_TA(2018)0337, (COM(2016)593. At http://www.europarl.europa.eu/doceo/document/TA-8-2018-0337_EN.pdf?redirect

Evans, D. S. and Schmalensee, R. (2016) *Matchmakers: The new economics of multisided platforms.* Boston, MA: Harvard University Press.

Facebook. (n.d.) 'Our mission', *Facebook.* At https://about.fb.com/company-info/

FCC. (1980) *FCC Computer Inquiry II final decision.* In the Matter of Amendment of Section 64.702 of the Commission's Rules and Regulations (Second Computer Inquiry). Docket No. 20828, Federal Communications Commission. Washington, DC. At http://etler.com/FCC/pdf/Numbered/20828/FCC%2080-189.pdf

Feld, H. (2019) *The case for the Digital Platform Act: Market structure and regulation of digital platforms.* Roosevelt Institute.Org and Publicknowledge.Org Paper. At https://rooseveltinstitute.org/the-case-for-the-digital-platform-act/

Fenton, N. (2019) '(Dis)trust'. *Journalism, 20*(1): 36–39.

Flaxman, S., Goel, S. and Rao, J. M. (2016) 'Filter bubbles, echo chambers, and online news consumption'. *Public Opinion Quarterly, 80*(1): 298–320.

Flecker, J. (ed.) (2016) *Space, place and global digital work.* London: Palgrave Macmillan.

Flew, T., Martin, F. and Suzor, N. (2019) 'Internet regulation as media policy: Rethinking the question of digital'. *Journal of Digital Media & Policy, 10*(1): 33–50.

Floridi, L. (2019) 'Translating principles into practices of digital ethics: Five risks of being unethical'. *Philosophy & Technology, 32*(2): 185–193.

Foroohar, R. (2019) 'Big tech must pay for access to America's "digital oil"', *Financial Times,* 7 April. At https://www.ft.com/content/fd3d885c-579d-11e9-a3db-1fe89bedc16e?desktop=true&segmentId=d8d3e364-5197-20eb-17cf-2437841d178a

Freedman, D. (2008) *The politics of media policy.* Cambridge: Polity Press.

Freedman, D. (2015) 'Paradigms of media power'. *Communication, Culture & Critique, 8*(2): 273–289.

Freedman, D. (2019) '"Public service" and the journalism crisis: Is the BBC the answer?' *Television & New Media, 20*(3): 203–218.

Freeman, C. (1988) 'Introduction'. In G. Dosi, C. Freeman, R. Nelson, G. Silverberg and L. Soete (eds). *Technical change and economic theory* (pp. 1–8). London: Pinter Publishers.

Freeman, C. and Perez, C. (1988) 'Structural crises of adjustment, business cycles and investment behaviour'. In G. Dosi, C. Freeman, R. Nelson, G. Silverberg and L. Soete (eds). *Technical change and economic theory* (pp. 38–66). London: Pinter Publishers.

Freeman, C. and Soete, L. (1994) 'The biggest technological juggernaut that ever rolled: Information and communication technology (ICT) and its employment effects'. In *Work for all or mass unemployment? Computerised technical change into the 21st century* (pp. 39–66). London: Pinter Publishers.

Frey, C. B. (2019) *The technology trap: Capital, labor, and power in the age of automation*. Princeton, NJ: Princeton University Press.

Frey, C. B. and Osborne, M. A. (2017) 'The future of employment: How susceptible are jobs to computerization?' *Technological Forecasting & Social Change, 114*(January): 254–280.

Fuchs, C. (2015) 'Dallas Smythe today – The audience commodity, the digital labour debate, Marxist political economy and critical theory. Prolegomena to a digital labour theory of value'. *TripleC – Cognition, Communication and Co-operation, 10*(2): 692–740.

Fuller, D. B. (2016) *Paper tigers, hidden dragons: Firms and the political economy of China's technological development*. Oxford: Oxford University Press.

Funk, J. L. (2003) *Mobile disruption: The technologies and applications driving the mobile internet*. New York: Wiley-Interscience.

Galpaya, H., Perampalam, S. and Senanayake, L. (2018) 'Investigating the potential for micro-work and online-freelancing in Sri Lanka'. In L. Pupillo, E. Noam and L. Waverman (eds). *Digitized labor: The impact of the internet on employment* (pp. 229–250). London: Palgrave Macmillan.

Gandy, O. H. Jr. (1993) *The panoptic sort: A political economy of personal information*. Westview, CO: Westview Press.

Gangadharan, S. P. and Jedrzej, N. (2019) 'Decentering technolgy in discourse on discrimination'. *Information, Communication & Society, 22*(7): 882–899.

Gans, J. (2016) *The disruption dilemma*. Cambridge, MA: The MIT Press.

Garnham, N. (1997) 'Amartya Sen's "capabilities" approach to the evaluation of welfare: Its application to communications'. *Javnost-the Public, 4*(4): 25–34.

Gawer, A. (ed.) (2011) *Platforms, market and innovation*. Cheltenham, UK and Northampton, MA, USA: Edward Elgar Publishing.

Gawer, A. (2014) 'Bridging differing perspectives on technological platforms: Towards an integrative framework'. *Research Policy, 43*(7): 1239–1249.

Ghahramani, Z. (2015) 'Probabilistic machine learning and artificial intelligence'. *Nature, 521*(28 May): 424–459.

Gillespie, T. (2010) 'The politics of "platforms"'. *New Media & Society, 12*(3): 347–364.

Gillespie, T. (2018) *Custodians of the internet: Platforms, content moderation, and the hidden decisions that shape social media*. New Haven, CT: Yale University Press.

Goldfarb, A., Greenstein, S. M. and Tucker, C. E. (eds). (2015) *Economic analysis of the digital economy*. Chicago, IL: University of Chicago Press.

Google. (n.d.) 'Google cares deeply about journalism'. *Google News Initiative*. At https://newsinitiative.withgoogle.com/intl/en_gb/about/

Graham, M. (ed.) (2019) *Digital economies at global margins*. Cambridge, MA and Ottawa: The MIT Press with IDRC.

Graham, M. and Anwar, M. A. (2019) 'The global Gig Economy: Towards a planetary labour market?' *First Monday, 24*(4): 1–15.

Groenewegen, J., Spithoven, A. and van den Berg, A. (2010) *Institutional economics: An introduction.* London: Red Globe Press/Springer Nature.

Gurumurthy, A., Bharthur, D., Chami, N., with Vipra, J and Anwar, I. A. (2019) *Platform planet: Development in the intelligent economy.* IT for Change and IDRC, Dehli and Ottawa, June. At https://itforchange.net/platformpolitics/wp -content/uploads/2019/06/Platform-Planet-Development-in-the-Intelligence -Economy_ITfC2019.pdf

Hagiu, A. (2007) 'Merchant or two-sided platform?' *Review of Network Economies, 6*(2): 115–133.

Hagiu, A. and Wright, J. (2015a) 'Marketplace or reseller?' *Management Science, 61*(1): 184–203.

Hagiu, A. and Wright, J. (2015b) 'Multi-sided platforms'. *International Journal of Industrial Organization, 41*: 162–174.

Harris Interactive. (2019) *Adtech market research report.* Harris Interactive, UK, for Information Commissioner's Office and Ofcom. London. At https://www .ofcom.org.uk/__data/assets/pdf_file/0023/141683/ico-adtech-research.pdf

Heeks, R. (2018) *Information and communication technology for development (ICT4D).* London: Routledge.

Helberger, N. (2018) 'Challenging diversity – Social media platforms and a new conception of media diversity'. In M. Moore and D. Tambini (eds). *Digital dominance: The power of Google, Amazon, Facebook, and Apple* (pp. 153–175). Oxford: Oxford University Press.

Helberger, N., Kleinen-von Königslöw, K. and van der Noll, R. (2015) 'Regulating the new information intermediaries as gatekeepers of information diversity'. *Info: The Journal of Policy, Regulation and Strategy for Telecommunications, 17*(6): 50–71.

Helsper, E. J. (2017) 'The social relativity of digital exclusion: Applying relative deprivation theory to digital inequalities'. *Communication Theory, 27*(3): 223–242.

Hepp, A. (2016) 'Pioneer communities: Collective actors in deep mediatisation'. *Media, Culture & Society, 38*(6): 918–933.

Herzog, C., Hilker, H., Novy, L. and Torun, O. (eds). (2017) *Transparency and funding of public service media.* Wiesbaden: Springer VS.

Hesmondhalgh, D. (2017) 'Capitalism and the media: Moral economy, well being and capabilities'. *Media, Culture & Society, 39*(2): 202–218.

Hesmondhalgh, D. (2019a) *The cultural industries, 4th edition.* London: Sage.

Hesmondhalgh, D. (2019b) 'Have digital communication technologies democratized the media industries?' In J. Curran and D. Hesmondhalgh (eds). *Media and society, 6th edition* (pp. 101–120). London: Bloomsbury.

Hintz, A., Dencik, L. and Wahl-Jorgensen, K. (2019) *Digital citizenship in a datafied society.* Cambridge: Polity.

Hodgson, G. M. (1989) 'Institutional economic theory: The old versus the new'. *Review of Political Economy, 1*(3): 249–269.

Hoskins, G. T. (2019) 'Beyond "zero sum": The case for context in regulating zero rating in the global South'. *Internet Policy Review, 8*(1): 1–26.

IAB Europe. (2018) *AdEx benchmark 2018*. IAB Europe. At https://iabeurope.eu/wp -content/uploads/2019/07/IAB-Europe_AdEx-Benchmark-2018-Report_July-2019.pdf

Ibrus, I. (2010) Evolutionary dynamics of new media forms: The case of open mobile web. Unpublished PhD thesis, London School of Economics and Political Science. At http://etheses.lse.ac.uk/53/1/Ibrus_Evolutionary _Dynamics_of_New_Media_Forms.pdf

Ibrus, I. (ed.) (2019) *Emergence of cross-innovation systems: Audiovisual industries co-innovating with education, health care and tourism*. London: Emerald Insight.

IoT Security Foundation. (2018) *Understanding the contemporary use of vulnerability disclosure in consumer internet of things product companies*. At https:// www.iotsecurityfoundation.org/wp-content/uploads/2018/11/Vulnerability -Disclosure-Design-v4.pdf

Jia, L. and Winseck, D. (2018) 'The political economy of Chinese internet companies: Financialization, concentration, and capitalization'. *International Communication Gazette, 80*(1): 30–59.

Jie, S. (2018) 'China exports facial ID technology to Zimbabwe', *Global Times*, 4 December. At http://www.globaltimes.cn/content/1097747.shtml

Just, N. (2018) 'Governing online platforms: Competition policy in times of platformization'. *Telecommunications Policy, 42*(5): 386–394.

Kadar, M. and Bogdan, M. (2017) '"Big data" and EU merger control – A case review'. *Journal of European Competition Law & Practice, 8*(8): 479–491.

Kahneman, D. and Tversky, A. (1979) 'Prospect Theory: Analysis of decision under risk'. *Econometrica, 47*(2): 263–291.

Kant, I. (1785/2012) *Groundwork of the metaphysics of morals*, trans. M. Gregor and J. Timmerman. Cambridge: Cambridge University Press.

Katz, J. E. and Aakhus, M. (eds). (2002) *Perpetual contact: Mobile communication, private talk, public performance*. Cambridge: Cambridge University Press.

Keen, A. (2015) *The internet is not the answer*. New York: Atlantic Books.

Khan, L. M. (2017) 'Amazon's antitrust paradox'. *The Yale Law Journal, 126*(3): 710–883.

Kimball, M. S. (2015) *Cognitive economics*. NBER Working Paper 20834. Cambridge, MA. At https://www.nber.org/papers/w20834.pdf

Kleine, D. (2013) *Technologies of choice? ICTs, development, and the capabilities approach*. Cambridge, MA: The MIT Press.

Kretschmer, M., Jensen, M., Radmacher, M., Rey-Moreno, C. and Schutz, E. (2019) *Connecting the unconnected: Tackling the challenge of cost-effective broadband internet in rural areas*. Fraunhofer Institute for Applied Information Technology FIT. Sankt Augustin. At https://www.wiback.org/content/dam/ wiback/en/documents/Study_Connect%20the%20Unconnected_2019.pdf

LaGrandeur, K. and Hughes, J. J. (2017) *Surviving the Machine Age: Intelligent technology and the transformation of human work*. London: Palgrave Macmillan.

Lanier, J. and Weyl, E. G. (2018) 'A blueprint for a better digital society'. *Harvard Business Review, Digital Article* (26 September): n.p.

Lehr, W., Clark, D. D., Bauer, S. and Claffy, K. C. (2019) *Regulation when platforms are layered*. Paper prepared for TPRC47. Washington, DC. At https:// papers.ssrn.com/sol3/papers.cfm?abstract_id=3427499

Lemstra, W. and Melody, W. H. (eds). (2014) *The dynamics of broadband markets in Europe: Realizing the 2020 Digital Agenda*. Cambridge: Cambridge University Press.

Livingstone, S. and Wang, Y. (2011) *Media literacy and the Communications Act*. London School of Economic and Political Science, Media Policy Brief 2. At http://www.lse.ac.uk/media@lse/documents/MPP/LSEMPPBrief2.pdf

Livingstone, S., Ólafsson, K., Helsper, E. J., Lupiáñez-Villanueva, F., Veltri, G. A. and Folkvord, F. (2017) 'Maximizing opportunities and minimizing risks for children online: The role of digital skills in emerging strategies of parental mediation'. *Journal of Communication, 67*(1): 82–105.

Lowe, G. G., Van den Bulck, H. and Donders, K. (eds). (2017) *Public service media in the networked society*. Gothenburg: NORDICOM.

Lyon, D. (1994) *The electronic eye: The rise of surveillance society*. Cambridge: Polity Press.

Lyon, D. (2018) *The culture of surveillance: Watching as a way of life*. Cambridge: Polity Press.

MacKenzie, D. (1989) 'Technology and the arms race. Review: Innovation and the arms race: How the United States and Soviet Union develop new military technologies by Matthew Evangelista'. *International Security, 14*(1): 161–175.

Mansell, R. (1996) 'Designing networks to capture customers: Policy and regulatory issues for the new telecommunications'. In W. H. Melody (ed). *Telecom reform: Principles, policies and regulatory practices* (pp. 83–96). Lyngby: Den Private Ingeniorfond, Technical University of Denmark.

Mansell, R. (2002) 'From digital divides to digital entitlements in knowledge societies'. *Current Sociology, 50*(3): 407–426.

Mansell, R. (2010) 'The life and times of the information society'. *Prometheus, 28*(2): 165–186.

Mansell, R. (2012) *Imagining the internet: Communication, innovation and governance*. Oxford: Oxford University Press.

Mansell, R. (2017) 'The mediation of hope: Communication technologies and inequality in perspective'. *International Journal of Communication, 11*: 4285–4304.

Mansell, R., Avgerou, C., Quah, D. and Silverstone, R. (eds). (2007) *The Oxford handbook of information and communication technologies*. Oxford: Oxford University Press.

Manyika, J., Lund, S., Chui, M., Bughin, J., Woetzel, J., Batra, P., Ko, R. et al. (2017) *Jobs lost, jobs gained: What the future of work will mean for jobs, skills, and wages*. McKinsey Global Institute. At https://www.mckinsey.com/featured-insights/future-of-work/jobs-lost-jobs-gained-what-the-future-of-work-will-mean-for-jobs-skills-and-wages

Manyozo, L. (2017) *Communicating development with communities*. London: Routledge.

Marcolin, L., Miroudot, S. and Squicciarini, M. (2016) *Routine jobs, employment and technological innovation in global value chains*. STI Policy Note, OECD, Paris, February. At https://www.oecd.org/sti/ind/GVC-Jobs-Routine-Content-Occupations.pdf

Mari, A. (2019) 'BBC seeks to incease younger audiences through data analytics', *Computer Weekly*, 5 February. At https://www.computerweekly.com/news/252456977/BBC-seeks-to-increase-younger-audience-through-data-analytics

Marsden, C. (2017) *Network neutrality: From policy to law to regulation.* Manchester: Manchester University Press.

Marsden, C. (2018a) 'How law and computer science can work together to improve the information society'. *Communications of the ACM, 61*(1): 29–31.

Marsden, C. (2018b) 'Prosumer law and network platform regulation: The long view towards creating Offdata'. *Georgetown Law Technology Review, 2*(2): 376–398.

Mattelart, A. (2000) *Networking the world: 1794–2000.* Minneapolis: University of Minnesota Press.

Mayer-Schönberger, V. and Cukier, K. (2013) *Big data: A revolution that will transform how we live, work and think.* London: John Murray.

McAfee, A. and Brynjolfsson, E. (2017) *Machine, platform, crowd: Harnessing our digital future.* New York: W. W. Norton & Company.

McGuigan, L. (2019) 'Automating the audience commodity: The unacknowledged ancestry of programmatic advertising'. *New Media & Society, 21*(11): 2366–2385.

McGuigan, L. and Manzerolle, V. (2014) '"All the world's a shopping cart": Theorizing the political economy of ubiquitous media and markets'. *New Media & Society, 17*(11): 1830–1848.

Meng, B. (2018) *The politics of Chinese media: Consensus and contestation.* New York: Palgrave Macmillan.

Milan, S. (2019) 'Emancipatory communication'. In R. Hobbs and P. Mihailidis (eds). *The international encyclopedia of media literacy* (n.p.). Hoboken, NJ: Wiley Blackwell.

Mitchell, W. J. (1999) *E-topia: Urban life, Jim-But not as we know it.* Cambridge, MA: The MIT Press.

Moe, H. (2010) 'Governing public service broadcasting: "Public value tests" in different national contexts'. *Communication, Culture & Critique, 3*(2): 207–223.

Moore, M. (1995) *Creating public value: Strategic management in government.* Cambridge, MA: Harvard University Press.

Moreno-Rey, C. and Graaf, M. (2017) 'Map of the community network initiatives in Africa'. In L. Belli (ed). *Community networks: The internet by the people, for the people: Official outcome of the UN IGF dynamic coalition on community connectivity* (pp. 147–169). Geneva: Internet Governance Forum. At http://bibliotecadigital.fgv.br/dspace/bitstream/handle/10438/19401/Community%20networks%20-%20the%20Internet%20by%20the%20people,%20for%20the%20people.pdf?sequence=1&isAllowed=y

Morozov, E. (2010) 'Think again: The internet', *Foreign Policy*, 26 April. At https://foreignpolicy.com/2010/04/26/think-again-the-internet/

Morton, F. S., Bourvier, P., Ezrachi, A., Jullien, B., Katz, R., Kimmelman, G., Melamed, D. et al. (2019) *Report. Commmittee for the study of digital platforms. Market structure and antitrust subcommittee.* Stigler Center for the Study of the Economy and the State. At https://research.chicagobooth.edu/-/media/research/stigler/pdfs/market-structure-report.pdf?la=en&hash=E08C7C9AA7367F2D612DE24F814074BA43CAED8C

Mosco, V. (2014) *To the cloud: Big data in a turbulent world.* Boulder, CO: Paradigm Publishers.

Mosco, V. (2019) *The smart city in a digital world*. Bingley: Emerald Publishing Ltd.

Mueller, M. L. (2010) *Networks and states: The global politics of internet governance*. Cambridge, MA: The MIT Press.

Mueller, M. L. and Grindal, K. (2019) 'Data flows and the digital economy: Information as a mobile factor of production'. *Digital Policy, Regulation and Governance, 21*(1): 71–87.

Murphy, H. (2019) 'Facebook advertising revenue withstands controversies', *The Financial Times*, 31 October. At https://www.ft.com/content/f35d665e-fb3a-11e9-98fd-4d6c20050229

Musgrave, R. (1959) *The theory of public finance: A study in political economy*. New York: McGraw-Hill.

Musiani, F., Cogburn, D. L., DeNardis, L. and Levinson, N. S. (2016) *The turn to infrastructure in internet governance*. Basingstoke: Palgrave Macmillan.

Nelson, R. R. (ed.) (1962) *The rate and direction of inventive activity: Economic and social factors*. Princeton, NJ: Princeton University Press.

NESTA. (2019) *Data trusts: A new tool for data governance*. NESTA. London. At https://hello.elementai.com/rs/024-OAQ-547/images/Data_Trusts_EN_201914.pdf

Newman, N. and Fletcher, R. (2018) 'Platform reliance, information intermediaries, and news diversity'. In M. Moore and D. Tambini (eds). *Digital dominance: The power of Google, Amazon, Facebook and Apple* (pp. 133–152). Oxford: Oxford University Press.

Newman, N., with Fletcher, R., Kalogeropoulos, A. and Kleis Nielsen, R. (2019) *Reuters Institute digital news report 2019*. Reuters Institute for the Study of Journalism. Oxford. At https://reutersinstitute.politics.ox.ac.uk/sites/default/files/2019-06/DNR_2019_FINAL_0.pdf

Noam, E. (2019) 'Looking ahead at internet video and its societal impacts'. In M. Graham and W. H. Dutton (eds). *Society & the internet: How networks of information and communication are changing our lives, 2nd edition* (pp. 371–388). Oxford: Oxford University Press.

Nooren, P., van Gorp, N., van Eijk, N. and Fathaigh, R. Ó. (2018) 'Should we regulate digital platforms? A new framework for evaluating policy options'. *Policy & Internet, 10*(3): 264–301.

North, D. C. (1977) 'Markets and other allocation systems in history: The challenge of Karl Polanyi'. *Journal of European Economic History, 6*(3): 703–716.

North, D. C. (1990) *Institutions, institutional change and economic performance*. Cambridge: Cambridge University Press.

OECD. (2015) *In it together: Why less inequality benefits all*. OECD. Paris. At http://www.oecd.org/social/in-it-together-why-less-inequality-benefits-all-9789264235120-en.htm

OECD. (2018) *Bridging the digital gender divide: Include, upskill, innovate*. OECD. Paris. At http://www.oecd.org/internet/bridging-the-digital-gender-divide.pdf

OECD. (2019) *Secretariat proposal for a 'unified approach' under pillar one*. OECD Public Consultation Document. Paris. At http://www.oecd.org/tax/beps/public-consultation-document-secretariat-proposal-unified-approach-pillar-one.pdf

O'Neil, C. (2016) *Weapons of math destruction: How big data increases inequality and threatens democracy*. Cambridge, MA: Harvard University Press.

O'Neill, O. (2002) *A question of trust: The BBC Reith Lectures 2002*. Cambridge: Cambridge University Press.

Ostrom, E. (1990) *Governing the commons: The evolution of institutions for collective action*. Cambridge: Cambridge University Press.

Pariser, E. (2011) *The filter bubble: What the internet is hiding from you*. New York: Penguin.

Parks, L. and Starosielski, N. (2015) *Signal traffic: Critical studies of media infrastructures*. Urbana: University of Illinois Press.

Pasquale, F. (2019) 'A rule of persons, not machines: The limits of legal automation'. *The George Washington Law Review, 87*(1): 1–55.

Paul, K. (2019) '"Breathtaking arrogance": Senators grill Facebook in combative hearing over Libra currency', *The Guardian*, 16 July. At https://www.theguardian.com/technology/2019/jul/15/big-tech-behemoths-face-grilling-from-us-lawmakers-as-hearings-kick-off

Pickard, V. (2020) *Democracy without journalism: Confronting the misinformation society*. New York: Oxford University Press.

Pissarides, C. and Thomas, A. (2019) *The future of good work: The foundation of a modern moral economy*. Institute for the Future of Work (IFOW) Discussion Paper. London. At https://www.ifow.org/publications/2019/2/13/the-future-of-good-work-the-foundation-of-a-modern-moral-economy

Plantin, J.-C., Lagoze, C., Edwards, P. N. and Sandvig, C. (2018) 'Infrastructure studies meet platform studies in the age of Google and Facebook'. *New Media & Society, 20*(1): 293–310.

Pupillo, L., Noam, E. and Waverman, L. (eds). (2018) *Digitized labor: The impact of the internet on employment*. London: Palgrave Macmillan.

Puppis, M. and Winseck, D. (Compilers). (2019) *Platform regulation inquiries, reviews and proceedings worldwide*. Resource, University of Fribourg and Carleton University. At https://docs.google.com/document/d/1AZdh9sECGfTQEROQjo5fYeiY_gezdf_11B8mQFsuMfs/edit#heading=h.drjg9uyede6x

Raboy, M. (2016) *Marconi: The man who networked the world*. Oxford: Oxford University Press.

Ranking Digital Rights. (2019) *Corporate Accountability Index*. New America Foundation. Washington, DC. At https://rankingdigitalrights.org/

Ren, J., Jubois, D. J., Choffnes, D., Mandalari, A. M., Kolcun, R. and Haddadi, H. (2019) *Information exposure from consumer IoT devices*. Proceedings of the Internet Measurement Conference IMC'19. New York.

Rheingold, H. (2000) *The virtual community: Homesteading on the electronic frontier* (Revised edition). Cambridge, MA: The MIT Press.

Ricaurte, P. (2019) 'Data epistemologies, the coloniality of power, and resistance'. *Television & New Media, 20*(4): 350–365.

Richards, N. and Hartzog, W. (2019) 'The pathologies of digital consent'. *Washington University Law Reviw, 96*: 1461–1503.

Roberts, H., Cowls, J., Morley, J., Taddeo, M., Wang, V. and Floridi, L. (2019) *The Chinese approach to artificial intelligence: An analysis of policy and regulation*. University of Oxford. At https://papers.ssrn.com/sol3/papers.cfm?abstract_id=3469784

Robinson, L., Cotten, S. R., Ono, H., Quan-Haase, A., Mesch, G., Chen, W., Schulz, J. et al. (2015) 'Digital inequalities and why they matter'. *Information, Communication and Society, 18*(5): 569–582.

Rochet, J.-C. and Tirole, J. (2003) 'Platform competition in two-sided markets'. *Journal of the European Economic Association, 1*(4): 990–1029.

Rohlfs, J. (1974) 'A theory of interdependent demand for a communications service'. *Bell Journal of Economics, 5*(1): 16–37.

Romanosky, J. and Chetty, M. (2018) 'Understanding the use and impact of the zero-rated Free Basics platform in South Africa'. *CHI'18 Proceedings of the 2018 CHI Conference on Human Factors in Computing Systems, Paper No. 192*: 1–13.

Romer, P. (1990) 'Endogenous technological change'. *Journal of Political Economy, 98*(5 Pt 2): 71–102.

Rutherford, M. (1994) *Institutions in economics: The old and the new institutionalism.* Cambridge: Cambridge University Press.

Samarajiva, R. and Mukherjee, R. (1991) 'Regulation of 976 services and dial-a-porn: Implications for the intelligent network'. *Telecommunications Policy, 15*(2): 151–164.

Samuelson, P. A. (1954) 'The pure theory of public expenditure'. *Review of Economics and Statistics, 36*(4): 387–389.

Schiller, H. and Miège, B. (1990) 'Communication of knowledge in an information society'. In J. Berleur, A. Clement, R. Sizer and D. Whitehouse (eds). *The information society: Evolving landscapes* (pp. 161–167). Concord, ON: Captus Press.

Schlosberg, J. (2018) 'Digital agenda setting: Reexamining the role of platform monopolies'. In M. Moore and D. Tambini (eds). *Digital dominance: The power of Google, Amazon, Facebook and Apple* (pp. 202–218). Oxford: Oxford University Press.

Schwab, K. (2017) *The fourth industrial revolution.* Geneva: World Economic Forum.

Schwab, K. (ed.). (2019) *The global competitiveness report 2019.* Geneva: World Economic Forum Insight Report.

ScrapeHero. (2019) 'How many products does Amazon sell worldwide – January 2018', *ScrapeHero.* At https://www.scrapehero.com/how-many-products-amazon-sell-worldwide-january-2018/

Sedgewick, R. and Wayne, K. (2011) *Algorithms, 4th edition.* Boston, MA: Addison Wesley Professional.

Selbst, A. D., Boyd, D., Friedler, S., Venkatasubramanian, S. and Vertesi, J. (2019) 'Fairness and abstraction in sociotechnical systems'. *FAT* 19 Proceedings of the Conference on Fairness, Accountability, and Transparency, 1*(1): 59–68.

Shen, H. (2016) 'China and global internet governance: Toward an alternative analytical framework'. *Chinese Journal of Communication, 9*(3): 304–324.

Shiller, R. J. (2019) *Narrative economics: How stories go viral & drive major economic events.* Princeton, NJ: Princeton University Press.

Siepel, J., Camerani, R., Pellegrino, G. and Masucci, M. (2016) *The fusion effect: The economic returns to combining arts and science skills.* NESTA. London. At https://media.nesta.org.uk/documents/the_fusion_effect_v6.pdf

Simon, H. A. (1991) *Models of my life.* New York: Basic Books.

Simonite, T. (2018). AI has started cleaning up Facebook, but can it finish? *Wired*, 18 December. At https://www.wired.com/story/ai-has-started-cleaning -facebook-can-it-finish/

Soete, L. (1985) 'International diffusion of technology, industrial development and technological leapfrogging'. *World Development, 13*: 409–422.

Soete, L. and Kamp, K. (1996) 'The "bit tax": The case for further research'. *Science and Public Policy, 23*(6): 353–360.

Solomonoff, G. (1956) *Ray Solomonoff and the Dartmouth Summer Research Project in artificial intelligence.* Oxbridge Research. Cambridge, MA. At http:// raysolomonoff.com/dartmouth/dartray.pdf

Souter, D. and Van der Spuy, A. (2019) *UNESCO's Internet Universality Indicators.* UNESCO. Paris. At https://unesdoc.unesco.org/ark:/48223/pf0000367617

Spithoven, A. (2019) 'Similarities and dissimilarities between original institutional economics and new institutional economics'. *Journal of Economic Issues, 53*(2): 440–447.

Srnicek, N. (2017) *Platform capitalism.* Cambridge: Polity.

Statista. (2019) 'Global no. 1 business data platform'. At https://www.statista.com/

Steinmueller, W. E. (2000) 'Will new information and communication technologies improve the "codification" of knowledge?' *Industrial and Corporate Change, 9*(2): 361.

Steinmueller, W. E. (2001) 'ICTs and the possibilities of leapfrogging by developing countries'. *International Labour Review, 140*(2): 193–210.

Steinmueller, W. E. (2003) *Information society consequences of expanding the intellectual property domain.* STAR Issue Report No. 39, SPRU, University of Sussex. London. At https://www.dime-eu.org/files/active/1/steinmueller.pdf

Stirling, A. (2007) 'A general framework for analysing diversity in science, technology and society'. *Journal of the Royal Society Interface, 4*(15): 707–719.

Stoilova, M., Rishita, N. and Livingstone, S. (2019) 'Children's understanding of personal data and privacy online: A systematic evidence mapping'. *Information, Communication & Society.* Published Online 17 September.

Strassburg, B. (1970) Address to the communications carriers and management information systems by the Chief of FCC Common Carrier Bureau, at Institute on Management Information and Data Transfer Systems, American University. Cybertelecom – Federal Internet Law & Policy, An Educational Project. Washington, DC. At https://www.cybertelecom.org/ci/ci.htm

Sullins, J. P. (2016) 'Ethics boards for research in robotics and artificial intelligence: Is it too soon to act?'. In M. Norskov (ed). *Social robots: Boundaries, potential, challenges* (pp. 83–98). Farnham: Ashgate Publishing.

Sunstein, C. R. (2009) *Republic.Com 2.0.* Princeton, NJ: Princeton University Press.

Taddeo, M. and Floridi, L. (2018) 'Regulate artificial intelligence to avert cyber arms race'. *Nature, 556*(19 April): 296–298.

Taye, B. and Cheng, S. (2019) 'The state of internet shutdowns', *Accessnow*, 8 July. At https://www.accessnow.org/the-state-of-internet-shutdowns-in-2018/.

Thaler, R. H. and Sunstein, C. R. (2009) *Nudge: Improving decisions about health, wealth, and happiness.* London: Penguin Books.

The Royal Society. (2018) *The malicious use of artificial intelligence: Forecasting, prevention, and mitigation.* The Royal Society. London. At https://arxiv.org/pdf/1802.07228.pdf.

Tirole, J. (2017) *Economics for the common good,* trans. S. Rendall. Princeton, NJ: Princeton University Press.

Trust Truth and Technology Commission. (2018) *Tackling the information crisis: A policy framework for media system resilience.* Report of the LSE Commission on Truth, Trust and Technology. London. At http://www.lse.ac.uk/media-and-communications/truth-trust-and-technology-commission/The-report.

Turow, J. (2011) *The daily you: How the new advertising industry is defining your identity and your worth.* New Haven, CT: Yale University Press.

Tversky, A. and Kahneman, D. (1992) 'Advances in prospect theory: Cumulative representation of uncertainty'. *Journal of Risk and Uncertainty,* 5(4): 297–323.

UK. (2018a) *Adults' media use and attitudes online.* Ofcom. London. At https://www.ofcom.org.uk/research-and-data/media-literacy-research/adults/adults-media-use-and-attitudes.

UK. (2018b) *Algorithms in decision-making.* House of Commons Science and Technology Committee Fourth Report of Session 2017–19. London. At https://publications.parliament.uk/pa/cm201719/cmselect/cmsctech/351/351.pdf.

UK. (2018c) *Democracy disrupted? Personal information and political influence.* Information Commissioner's Office. London. At https://ico.org.uk/media/action-weve-taken/2259369/democracy-disrupted-110718.pdf.

UK. (2018d) *The economic value of data: Discussion paper.* HM Treasury. At https://assets.publishing.service.gov.uk/government/uploads/system/uploads/attachment_data/file/731349/20180730_HMT_Discussion_Paper_-_The_Economic_Value_of_Data.pdf.

UK. (2019a) *Disinformation and 'fake news': Final report, Eighth Report of Session 2017–19.* House of Commons, Digital, Culture, Media and Sport Committee. London. At https://publications.parliament.uk/pa/cm201719/cmselect/cmcumeds/1791/1791.pdf.

UK. (2019b) *The future of public service media.* Ofcom. London. At https://www.ofcom.org.uk/__data/assets/pdf_file/0022/155155/future-public-service-media.pdf.

UK. (2019c) *Media nations: UK 2019.* Ofcom. London. At https://www.ofcom.org.uk/__data/assets/pdf_file/0019/160714/media-nations-2019-uk-report.pdf.

UK. (2019d) *Online Harms White Paper.* Secretary of State for Digital, Culture, Media & Sport and the Secretary of State for the Home Department. London. At https://www.gov.uk/government/consultations/online-harms-white-paper.

UK. (2019e) *Online market failures and harms: An economic perspective on the challenges and opportunities in regulating online services.* Ofcom. London. At https://www.ofcom.org.uk/__data/assets/pdf_file/0025/174634/online-market-failures-and-harms.pdf.

UK. (2019f) *Public service broadcasting: As vital as ever.* House of Lords Select Committee on Communications and Digital, First Report of Session 2019. London. At https://publications.parliament.uk/pa/ld201919/ldselect/ldcomuni/16/16.pdf.

UK. (2019g) *Regulating in a digital world.* House of Lords Select Committee on Communications, Second Report of Session 2017–19. London. At https://publications.parliament.uk/pa/ld201719/ldselect/ldcomuni/299/299.pdf.

UK. (2019h) *The right to privacy (Article 8) and the digital revolution.* Joint Committee on Human Rights (HC 122 HL Paper 14). London. At https://publications.parliament.uk/pa/jt201919/jtselect/jtrights/122/122.pdf.

UNCTAD. (2019) *Digital Economy report 2019: Value creation and capture – implications for developing countries.* United Nations Conference on Trade and Development. Geneva. At https://unctad.org/en/pages/PublicationWebflyer.aspx?publicationid=2466.

US. (1990) *Critical connections: Communications for the future.* Washington, DC: Office of Technology Assessment.

US. (1996) *Communication Decency Act.* US Public Law 104–104, 8 February. Title V 47 U.S.C. S 230. At https://www.congress.gov/104/plaws/publ104/PLAW-104publ104.pdf.

US. (1998a) *Digital Millennium Copyright Act.* US Public Law 105–304, 28 October. United States. At https://www.govinfo.gov/content/pkg/PLAW-105publ304/pdf/PLAW-105publ304.pdf.

US. (1998b) *Privacy online: A report to Congress.* Federal Trade Commission. Washington, DC. At https://www.ftc.gov/sites/default/files/documents/reports/privacy-online-report-congress/priv-23a.pdf.

US. (2018a) *California Consumer Privacy Act of 2018.* State of California Legislature, Assembly Bill No. 375, Chapter 55. At https://leginfo.legislature.ca.gov/faces/billTextClient.xhtml?bill_id=201720180AB375.

US. (2018b) *Charting a course for success: America's strategy for STEM education.* Executive Office of the President Report by Committee on STEM Education, National Science & Technology Council (NSTC). Washington, DC. At https://www.whitehouse.gov/wp-content/uploads/2018/12/STEM-Education-Strategic-Plan-2018.pdf.

US. (2019a) *Digital Citizenship and Media Literacy Act.* Introduced in Senate US Congress, 116th Congress 2019–2020. Washington, DC. At https://www.congress.gov/bill/116th-congress/senate-bill/2240/text?r=8&s=1

US. (2019b) *Justice Department reviewing the practices of market-leading online platforms.* US Department of Justice. Washington, DC. At https://www.justice.gov/opa/pr/justice-department-reviewing-practices-market-leading-online-platforms

Ustek-Spilda, F., Powell, A. and Nemorin, S. (2019) 'Engaging with ethics in Internet of Things: Imaginaries in the social milieu of technology developers'. *Big Data & Society*, July–December: 1–12.

van Dijck, J., Poell, T. and De Waal, M. (2018) *The platform society: Public values in a connective world.* Oxford: Oxford University Press.

van Dijk, J. A. G. M. (2013) 'A theory of the digital divide'. In M. Ragnedda and G. W. Muschert (eds). *The digital divide: The internet and social inequality in international perspective* (pp. 29–51). New York: Routledge.

Van Doorn, N. (2017) 'Platform labor: On the gendered and racialized exploitation of low-income service work in the "on-demand" economy'. *Information, Communication & Society, 20*(6): 898–914.

van Schewick, B. (2010) *Internet architecture and innovation.* Cambridge, MA: The MIT Press.

Varian, H. R. (2016) 'The economics of internet search'. In J. M. Bauer and M. Latzer (eds). *Handbook on the economics of the internet* (pp. 385–394). Cheltenham, UK and Northampton, MA, USA: Edward Elgar Publishing.

WEF. (2017) '8 men have the same wealth as 3.6 billion of the world's poorest people. We must rebalance this unjust economy'. World Economic Forum. At https://www.weforum.org/agenda/2017/01/eight-men-have-the-same-wealth-as-3-6-billion-of-the-worlds-poorest-people-we-must-rebalance-this-unjust-economy

WEF. (2018) *The future of jobs report 2018*. World Economic Forum, Centre for the New Economy and Society. At http://www3.weforum.org/docs/WEF_Future_of_Jobs_2018.pdf

WEF. (2019) *Agenda in focus: Fixing equality*. World Economic Forum based on Oxfam. At https://www.weforum.org/focus/fixing-inequality

Williamson, O. E. (1975) *Markets and hierarchies: Analysis and antitrust implications*. New York: Free Press.

Williamson, O. E. (2000) 'The new institutional economics: Talking stock, looking ahead'. *Journal of Economic Literature, 38*(3): 595–613.

Winseck, D. (2016) 'Reconstructing the political economy of communication for the digital media age'. *The Political Economy of Communication, 4*(4): 73–114.

Wu, T. (2016) *The attention merchants: The epic scramble to get inside our heads*. New York: Random House.

Wu, T. (2018) *The curse of bigness: Antitrust in the new Gilded Age*. New York: Penguin Random House.

Wu, T. (2019) 'Blind spot: The attention economy and the law'. *Antitrust Law Journal, 82*(3): 771–806.

Yang, F. and Mueller, M. L. (2019) 'Internet governance in contemporary China'. In J. Yu and S. Guo (eds). *The Palgrave handbook of local governance in contemporary China* (pp. 441–463). Singapore: Palgrave Macmillan.

Yang, G. (2017) 'Demobilizing the emotions of online activism in China: A civilizing process'. *International Journal of Communication, 11*: 1945–1965.

Yu, H. (2017) *Networking China: The digital transformation of the Chinese economy*. Chicago: University of Illinois Press.

Zelizer, B. (2017) *What journalism could be*. Polity Press. Cambridge.

Zittrain, J. (2003) 'Internet points of control'. *Boston College Law Review, 44*(2): 653–688.

Zuboff, S. (1988) *In the age of the smart machine: The future of work and power*. New York: Basic Books.

Zuboff, S. (2019) *The age of surveillance capitalism: The fight for a human future at the new frontier of power*. New York: Public Affairs.

Zuckerberg, M. (2019) 'A privacy-focused vision for social networking', *Facebook*, 6 March. At https://www.facebook.com/notes/mark-zuckerberg/a-privacy-focused-vision-for-social-networking/10156700570096634/.

Index

Environmental Impact Assessment
Angus Morrison Saunders

Comparative Constitutional Law
Second Edition
Mark Tushnet

National Innovation Systems
*Cristina Chaminade, Bengt-Åke
Lundvall and Shagufta Haneef*

Ecological Economics
Matthias Ruth

Private International Law and Procedure
Peter Hay

Freedom of Expression
Mark Tushnet

Law and Globalisation
Jaakko Husa

Regional Innovation Systems
*Bjørn T. Asheim, Arne Isaksen and
Michaela Trippl*

International Political Economy
Second Edition
Benjamin J. Cohen

International Tax Law
Second Edition
Reuven S. Avi-Yonah

Social Innovation
*Frank Moulaert and Diana
MacCallum*

The Creative City
Charles Landry

European Union Law
Jacques Ziller

Planning Theory
Robert A. Beauregard

Tourism Destination Management
Chris Ryan

International Investment Law
August Reinisch

Sustainable Tourism
David Weaver

Austrian School of Economics
Second Edition
Randall G. Holcombe

U.S. Criminal Procedure
Christopher Slobogin

Platform Economics
*Robin Mansell and W. Edward
Steinmueller*

Public Finance
Vito Tanzi

Feminist Economics
Joyce P. Jacobsen

Human Dignity and Law
James R. May and Erin Daly

Space Law
Frans G. von der Dunk

Legal Research Methods
Ernst Hirsch Ballin

National Accounting
John M. Hartwick

International Human Rights Law
Second Edition
Dinah L. Shelton

Privacy Law
Megan Richardson

Law and Artificial Intelligence
Woodrow Barfield and Ugo Pagello